Hypergraphics

Visualizing
Complex Relationships in
Art, Science and Technology

AAAS Selected Symposia Series

Routledge
Taylor & Francis Group

LONDON AND NEW YORK

Hypergraphics

Visualizing
Complex Relationships in
Art, Science and Technology

Edited by David W. Brisson

AAAS Selected Symposium 24

First published 1978 by Westview Press, Inc.

Published 2018 by Routledge
52 Vanderbilt Avenue, New York, NY 10017
2 Park Square, Milton Park, Abingdon, Oxon OX14 4RN

Routledge is an imprint of the Taylor & Francis Group, an informa business

Library of Congress Catalog Card Number: 78-19725
ISBN 13: 978-0-367-01820-7 (hbk)
ISBN 13: 978-0-367-16807-0 (pbk)

About the Book

Visualization in its many forms has been a major contributing factor in the historical development of scientific thought. Yet in recent years there has been an increasing tendency to exclude visualization from the process of scientific exploration, and a similar tendency to exclude abstract mathematical formulations from the arts. These dissociations are destructive to a full, rich comprehension of either the arts or the sciences. To counter this trend, hypergraphics — a new field comprising computer graphics, perceptual psychology, and modern geometry — endeavors to develop new aids to understanding and communicating complex, multidimensional relationships.

Hypergraphics blends contemporary thinking in art and science by developing methods whereby our ability to visualize may be brought into active cooperation with our ability to reason abstractly, thus integrating these two modes of human thought and consciousness. This volume reviews developments in hypergraphics from the perspectives of a variety of fields in art and science.

About the Series

The *AAAS Selected Symposia Series* was begun in 1977 to provide a means for more permanently recording and more widely disseminating some of the valuable material which is discussed at the AAAS Annual National Meetings. The volumes in this *Series* are based on symposia held at the Meetings which address topics of current and continuing significance, both within and among the sciences, and in the areas in which science and technology impact on public policy. The *Series* format is designed to provide for rapid dissemination of information, so the papers are not typeset but are reproduced directly from the camera-copy submitted by the authors, without copy editing. The papers are organized and edited by the symposium arrangers who then become the editors of the various volumes. Most papers published in this *Series* are original contributions which have not been previously published, although in some cases additional papers from other sources have been added by an editor to provide a more comprehensive view of a particular topic. Symposia may be reports of new research or reviews of established work, particularly work of an interdisciplinary nature, since the AAAS Annual Meetings typically embrace the full range of the sciences and their societal implications.

WILLIAM D. CAREY
Executive Officer
American Association for
the Advancement of Science

Contents

About the Editor and Authors

David W. Brisson, associate professor of design at the Rhode Island School of Design, is a painter and sculptor, and for the past 15 years has been working in the area of geometry and perception. He invented the three-dimensional anaglyph and the hyperstereopticon and devised a method of making three-dimensional lenticullar drawings, hyperstereograms, and other forms. His art has been exhibited throughout the United States, and he has organized a number of symposia and exhibitions on hypergraphics. Among his books are Curved 4-Space *(1976), and* A New Reality: A Theory of Dimensionality *(1976).*

Thomas F. Banchoff, professor of mathematics at Brown University, specializes in the geometry and topology of differential and polyhedral manifolds. He has received Woodrow Wilson and Danforth Fellowships and a Fulbright Travel Award and is a member of several professional societies. His numerous articles concern the geometry of curves and surfaces, polygons and polyhedra in 3- and 4-dimensional space, and computer graphics in geometric research.

Harriet E. Brisson, a sculptor and ceramicist at Rhode Island College, has shown her work in hypergraphics at various exhibitions, and her honors include first prize for sculpture at the Fall River Art Association National Exhibit and first prize from the Newport Art Association. She is also active in the American Crafts Council and was the U.S. delegate to the World Craft Council Conference in 1978.

Scott E. Kim, a computer programmer at Logicon, Inc., in San Pedro, California, holds a B.A. in music from Stanford University, where he worked at the Computer Center for Research in Music and Acoustics. He is interested in the contribution of computer graphics to mathematical and artistic visualization, particularly transformations in time. He is a member of SIGGRAPH, has written articles on mathematics and music, three- and four-dimensional dissections, and invertible calligraphy, and has contributed articles on tilings and tesseracts to the Scientific American *mathematical games column.*

C. Ernesto S. Lindgren, professor in the Department of Urban and Regional Planning at the Federal University of Rio de Janeiro, specializes in quantitative methods in urban and regional planning. He is the author of Four-Dimensional Descriptive Geometry *(with Steve M. Slaby; McGraw-Hill, 1968) and* Developments in Four-Dimensional Descriptive Geometry *(Brazil: COPPE/UFRJ, 1977) and is a recipient of the "Descriptive Geometry Award" from A.S.E.E.*

Arthur L. Loeb, senior lecturer in the Department of Visual and Environmental Studies at Harvard University, specializes in design science. He has published numerous articles on a broad range of subjects, including physics, structure and patterns, computers and programmed instruction, crystal algebra, and symmetry theory, and is the author of Color and Symmetry *(Wiley, 1971) and* Space Structures, Their Harmony and Counterpoint *(Addison-Wesley, 1976). His work has been exhibited at Harvard, Smith College, Rhode Island School of Design, and other locations, and he has been elected to numerous professional societies.*

A. Michael Noll, marketing supervisor at American Telephone and Telegraph Company, has worked extensively in computer graphics, man-machine communications, and speech processing. His computer art has been exhibited throughout the world and he has created several computer movies. A member of the editorial advisory board of Computers and Graphics, *he has published over 35 papers and been granted 5 patents in his fields of interest.*

Steve M. Slaby, associate professor in the Department of Civil Engineering at Princeton University, works primarily in descriptive geometry and the impact of technology on society. He has received awards from Tau Beta Pi and the Lawrence Institute of Technology. His publications include Fundamentals of Three-Dimensional Descriptive Geometry *(Harcourt, Brace & World, 1966) and* Four-Dimensional Descriptive Geometry *(with C.E. Lindgren; McGraw-Hill, 1968).*

Cyril Stanley Smith is Institute Professor Emeritus at the Massachusetts Institute of Technology. His areas of specialization are metallurgy and the history of technology, and he was director of the Institute for the Study of Metals at the University of Chicago for 15 years. He is the author of A History of Metallography *(University of Chicago Press,*

1960). He has been awarded honorary degrees from Case-Western University and the University of Pennsylvania.

Charles M. Strauss is a research associate in the Department of Mathematics at Brown University. He has worked on the use of computer graphics in geometric research and has developed methods for producing computer-generated films.

Anne Griswold Tyng is an independent architect and a lecturer at the Department of Architecture, Graduate School of Fine Arts, University of Pennsylvania. Her work is focused on geometric ordering principles at all levels of form, and she has published articles and video films on this topic and in related areas. She has lectured and exhibited her work worldwide, and she is a fellow of the American Institute of Architects and an associate member of the National Academy of Design.

J. M. Yturralde, an internationally known painter from Valencia, Spain, works also with computer-generated art forms, and has held several exhibitions and lectured on this subject in Europe. A Scholar of the Direccion General de Bellas Artes and of the Computer Center of the University of Madrid, he was awarded first prize in the International Ibizagrafic 72 Exhibition, as well as the international B.J. Salvi prize, and the Premio Europa Ancona Italy. He was a fellow at the Center for Advanced Visual Studies at MIT in 1975.

Hypergraphics

Visualizing
Complex Relationships in
Art, Science and Technology

Introduction

David W. Brisson

"Hypergraphics" may be described as graphics that transcend traditional means. In a very specific sense it refers
to n-dimensional descriptive geometry. In a broader sense it
refers to any transcendent visual concern. It is in this
broader sense that this collection of essays has been assembled.

The authors of these essays have been interacting with
each other for a number of years at symposia and exhibitions
and through the personal exchange of work and ideas. Since
the authors represent a broad spectrum of disciplines in the
arts and sciences, a superficial perusal of the titles of
these papers does not immediately give the reader a clear
sense of the fundamental core that binds them together. Upon
examining the material more closely it will be found that
these ties consist of a preoccupation with the visual, and
basically geometrical, transformation of information; the
novel character, beauty and complexity of such transformations; the concern with the promise of new methods for dealing with the many problems of the day; and finally, the
adventure of exploration.

Steve Slaby's work represents the most direct concern
with technical aspects of transcendent descriptive geometry.
He has been involved with the translation of an early German
text on the subject of four-dimensional geometry, an original
text on the subject in collaboration with C. Ernesto Lindgren, as well as other works. Steve's paper serves to supply
perspective on the subject.

C. Ernesto Lindgren has been involved with the application of such concepts to city planning and architecture for
many years. Ernesto and I had long conversations about this
potential of n-dimensional geometry ten years ago when he was
working with computor graphics at Harvard University. Since

1

that time he has continued, in Brazil, to develop such methods of application in conjunction with practical problems in
urban planning.

Arthur Loeb is a scientist who has long been interested
in the teaching of design. Through geometrical reasoning, he
attempts in his chapter to formulate a more rigorous approach
to the process of "visual" design, a "design science." His
teaching involves the extensive use of mathematical models,
and his encouragement of his students has produced superla-
tive results in the exploration of design by means of its
geometrical aspects. His work in this area extends beyond
the classroom in his active and vigorous support of an interdisciplinary group of scholars called the Philomorphs, who
are interested in visual aspects of form and who meet once a
month on the Harvard campus.

Several years ago, Arthur asked me to give a presentation on four-dimensional geometry to the Philomorphs. The
presentation evoked sufficient interest to generate an exhibition, "Virtual Realities," organized by Toshihiro Katayama
and held at the Carpenter Center for the Visual Arts at
Harvard in 1976. The participants in that exhibition included myself, Harriet Brisson, who was engaged at the time
in building tensegrity models of close-packing polyhedra, and
Jose Yturralde, the Spanish painter and graphic artist, who
was at that time a fellow at the Center for Advanced Visual
Studies at M.I.T. Jose's work consisted of silk-screen
prints of "impossible" geometrical figures (illustrated in
his paper in this volume).

Jose's paper is concerned with his attitude of exploration into a transcendant visual world and his philosophical
approach to his art. His work extends to three-demensional
models, and he has developed beautiful kites as an extension
of that work. (In a desire to participate more fully in this
work he practiced hang-gliding!) His home in Valencia is not
too far from Granada, site of the Alhambra, the great source
of geometrical art for many European artists, including
Escher to whom Jose owes much inspiration.

Harriet Brisson's paper, like Jose's, is in effect a
statement of personal philosophical objectives, an outline of
the concerns of her work. It is in her neon sculptures that
Harriet's work truly begins to transcend ordinary sculptural
form. Unfortunately, the photographs of her work do not
adequately give the visual "magic" of the semi-mirrored
images with their infinite extension and their glowing color.

My work in the "Virtual Realities" exhibition consisted of a variety of four-dimensional models in various media, and some hyperanaglyphs of several four-dimensional forms. The hyperanaglyph is an analog extension of the red and blue "glasses" used in 3-D comic books years ago. In this case a four-dimensional polytope is projected from four dimensions to three, constructed out of welded rods and painted red. The polytope is then rotated about six degrees in four dimensions around a plane, projected to three dimensions, constructed and painted blue. This compound structure is then rotated in three dimensions while being viewed with the red and blue glasses. The effects are rather startling and unlike any ordinary experience of three dimensions.

At the "Virtual Realities" exhibition Tom Banchoff and Charles Strauss of Brown University showed their films of rotating hyperfigures. The effects of these films are very similar to the effects of the hyperanaglyphs. They are remarkably beautiful and the reproductions in this volume, which are selected frames from their films, again do not do justice to their beautiful color and the incredible effects of the animation.

Tom and Charles have been actively engaged in trying to mesh the efforts of art students and mathematics students from the two schools by offering joint courses and encouraging interdisciplinary study. Tom and Charles form a close to perfect symbiosis, Tom supplying the geometrical theory and Charles the computer technology. Together, after long and laborious experimentation with hardware and software techniques, they have arrived at methods for producing computer-generated films that would appear to rival any computer-generated films produced anywhere, both in form and content.

These films are of course related directly to the pioneer work of A. Michael Noll whose 1967 essay is reprinted in this volume. Michael has been interested in computer graphics and transcendant images for a long time and has lectured widely on this subject. Michael's paper is one of several that he has written on computer graphics. One of his concerns, shared by Banchoff and me, has been to discover some means of perceiving, either directly or indirectly, a four-dimensional figure. His interest in this specific problem is but one of a wide range of computer-based concerns, which include his computer simulation of the neo-plastic work of the painter Piet Mondrian. He has been concerned with computer graphics as an art form in itself, and he has experimented with computer-generated tactile visualization, what Huxley would have called "feelies."

As a result of the Carpenter Center exhibition I was encouraged to organize an exhibition and symposium, "Hypergraphics," in 1977 at the Rhode Island School of Design. About forty scientists and artists participated in the exhibition and discussions. Among them were Arthur Loeb and Cyril S. Smith, who in his long, distinguished career as a scientist has maintained a deep commitment to art and its relation to science. Concurrent with the symposium, Cyril Smith developed an exhibition at the Smithsonian concerned with the technology of many art objects in history. His background as a metallurgist has given him a profound respect for geometry as one of the bases of organized thought, and Arthur Loeb's background in crystallography has similarly affected him. Cyril's work is careful and scholarly, and concerned with the unification of work in the arts and sciences. His concern revolves around the profound connection between visual and verbal thought in art and science which is the root of this whole study.

At this symposium Banchoff's and Strauss' films were again shown, this time to an enthusiastic audience of more than 200 art students. I showed my hyperanaglyphs and a somewhat crude version of a kinetic hyperstereopticon. Toshihiro Katayama exhibited a number of silk-screen prints of "impossible" figures and a number of other artists and scientists exhibited work. Another scientist who met with the group at that time was Gregg Edwards of the National Science Foundation who presented films produced over the last several years that were related to our concerns, and he discussed the interconnections between the various disciplines and the future implications of activities related to hypergraphics. He brought to the group a particularly wide range of interests and the desire to formulate significant generalizations.

Among the exhibitors was Denis Finch, a scientist turned sculptor, who exhibited kinetic shadowgraph sculptures of rotating hyperfigures. Beautifully crafted, they formed a connecting link between the films of Banchoff and Strauss and my hyperanaglyphs. Harriet Brisson exhibited new tensegrity close-packing structures, and Arthur Loeb exhibited models of the dissection of the cube. There were of course many others who exhibited work and participated in the discussions and who are still actively involved in the production of transcendant images. "Hypergraphics" was through their efforts and interplay quite a successful affair in many ways.

Following this exhibition and symposium, Gregg Edwards urged me to arrange a similar exhibition and symposium at the annual conference of the American Association for the Ad-

vancement of Science in Washington, D.C., in 1978. This was
duly held in February, and this set of essays is the result
of that meeting (although not all of the papers were actually
given there).

The "Hypergraphics" exhibition in Washington contained
prints of "impossible" figures by Katayama and Yturralde as
well as shadowgraphs by Finch, a neon sculpture by Harriet
Brisson and a hyperanaglyph and a number of watercolor hyperstereograms by myself. Several other artists, including
Naoki Yoshimoto of Tokyo, were intended to be included, but
the February blizzard that nearly paralyzed New England
prevented this work from arriving in Washington. In the
symposium itself Banchoff's and Strauss' films were again
received by an enthusiastic audience. I presented the hyperstereograms in polarized form for the first time, and Gregg
Edwards presented clips from some fresh, highly sophisticated
computer graphics.

Several newcomers to this nucleus of hypergraphics
enthusiasts joined us about that time, including architect
Anne Griswold Tyng. Anne has been writing for some years in
a highly imaginative and intellectually provocative way,
drawing connections between classical and esoteric architectural history, and she has a deep interest in the basic
geometrical aspects of form in architecture.

Also included is Scott Kim, a young mathematician from
Stanford who had seen some of my work at Martin Gardner's
office. He sent me the delightful paper on the four-dimensional analog of the "impossible" triangle. The relation of
this work to that of Yturralde is obvious, and clearly it
forms a connecting link between that work and others more
directly related to the classical aspects of descriptive
geometry.

As should be clear from the above, many of the writers
of this volume have known each other for a long time. In
spite of their differences with regard to professional credentials, they are perhaps closer to each other than they are
to their immediate professional colleagues.

The enthusiasm of this bond has its roots in the "fron-
tier" character of hypergraphics and the fact that it is
deeply concerned with visual/verbal transformations. "Seeing" is a tangible means of "understanding" something, and
hypergraphics offers methods of "seeing" what has been considered non-visualizable by many people for a long time.

In <u>Hypergraphics</u> I have collected a great many pieces

of information, some stretching back into the last century
and further, that have existed mostly as curiosities. I hope
that the authors' syntheses offer fresh means of considering
many of the technical, artistic, scientific and philosophical
concerns of the present day.

Communication between different disciplines is notori-
ously difficult. However, even though the descriptive geometry of n-dimensions is mechanically complex, the derived
images are often so dramatically powerful that, with proper
preparation, many very, very complicated mathematical concep-
tions that require an enormous amount of experience with
symbolic languages may be expressed to the nonmathematically
trained with a very high level of communicative impact, with
minimal effort on the part of the participant.

The contents of this volume are as deceptively hetero-
geneous in their form as the authors are diverse in their
professional credentials. Some are highly technical, and
perhaps only legible to the mathematically trained, while
some perhaps will appeal most to the visual designer. However, throughout these essays the concern with visual models,
visual thought, and the concept of the analog and the transformation is a persistent, pervasive thread, and those readers who wish to consider the relation of imagery to abstrac-
tion will find here things to delight them and food for the
generation of further elaboration of concepts and means that
are only briefly touched upon in these essays.

Geometry in Applied Science and Engineering

Steve M. Slaby

Geometry in the form of visual, descriptive, and graphic spatial relations has had a long historic development and which from all indications (as will be demonstrated by this session) will continue to evolve into the distant future.

If we consider that "geometry", as an ideal concept, is the unifying basis of physical matter and therefore of physical life itself, then we can speak of the "geometry of life" which manifests itself in many forms in living organisms. It is seen in the pentagonal symmetrys in living organisms, in harmonious physical growth, in the logarithmic spiral reflected in flowers and shell life, and in the order of atomic and sub-atomic structures. (See figs 1 & 2)

The graphical expressions which man has developed over the centuries has an evolutionary span ranging from the primitive drawings of the caveman, dating back to approximately 12,000 BC to the sophisticated and intricate complex multi-view, multi-dimensional, multi-colored graphical expressions of modern times including multi-motion computer graphics. This evolutionary development over the ages, points to the fact that in order to understand the natural-physical reality in which man is embedded, and of which he is an integral part, man has felt the need to create an "instrument" which makes it possible for him to visualize in his "minds eye" the truth of this reality. Man's eyes have looked at the physical reality which surrounds him but have not always "seen" nor fully understood what they have looked at. Artists, scientists, engineers, etc. therefore have used and represented their visions of reality (and abstractions of this reality) through the instrument of graphics which is grounded in the concept of the measurement of physical space geometries whether they be Egyptian pyramids, solar systems, industrial and technological systems, or atomic systems.

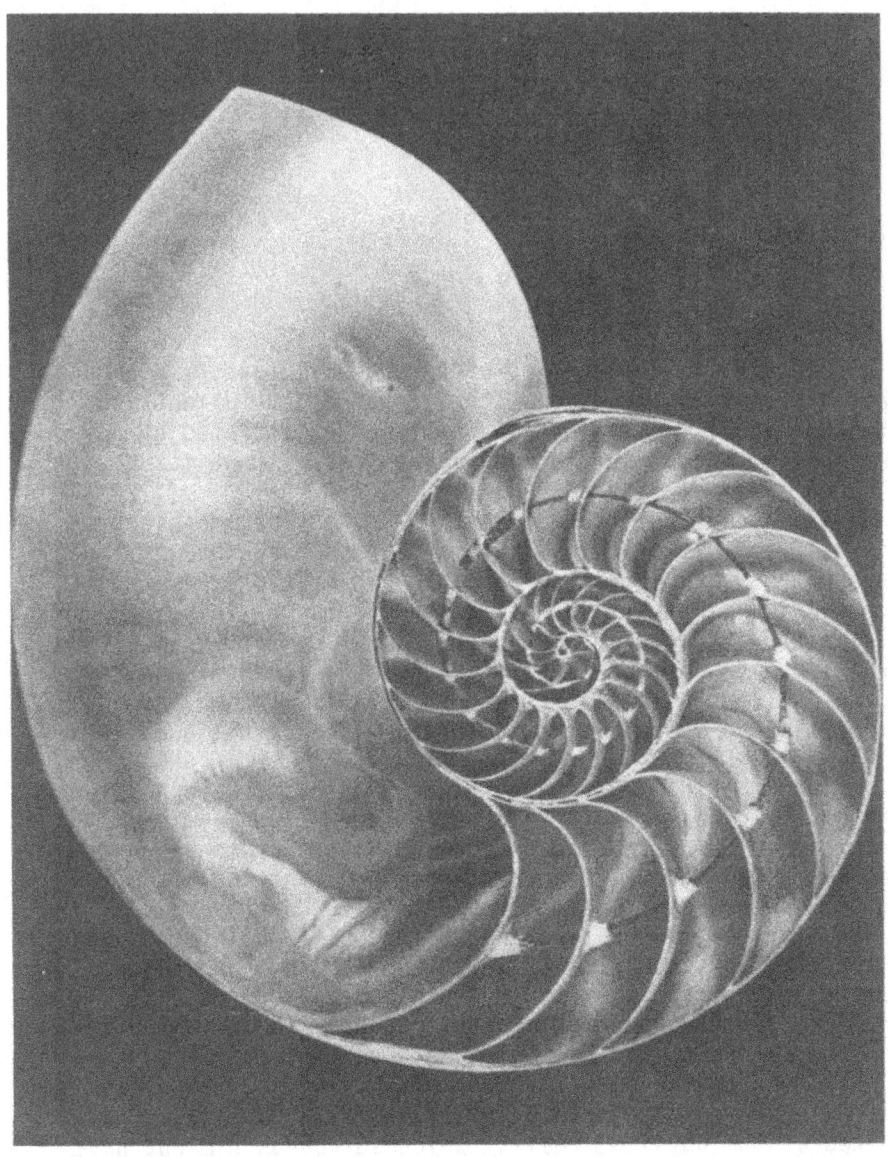

Figure 1

Source: "Mathematics", Life
Science Library, Time Inc.,
N.Y., 1963, p. 92

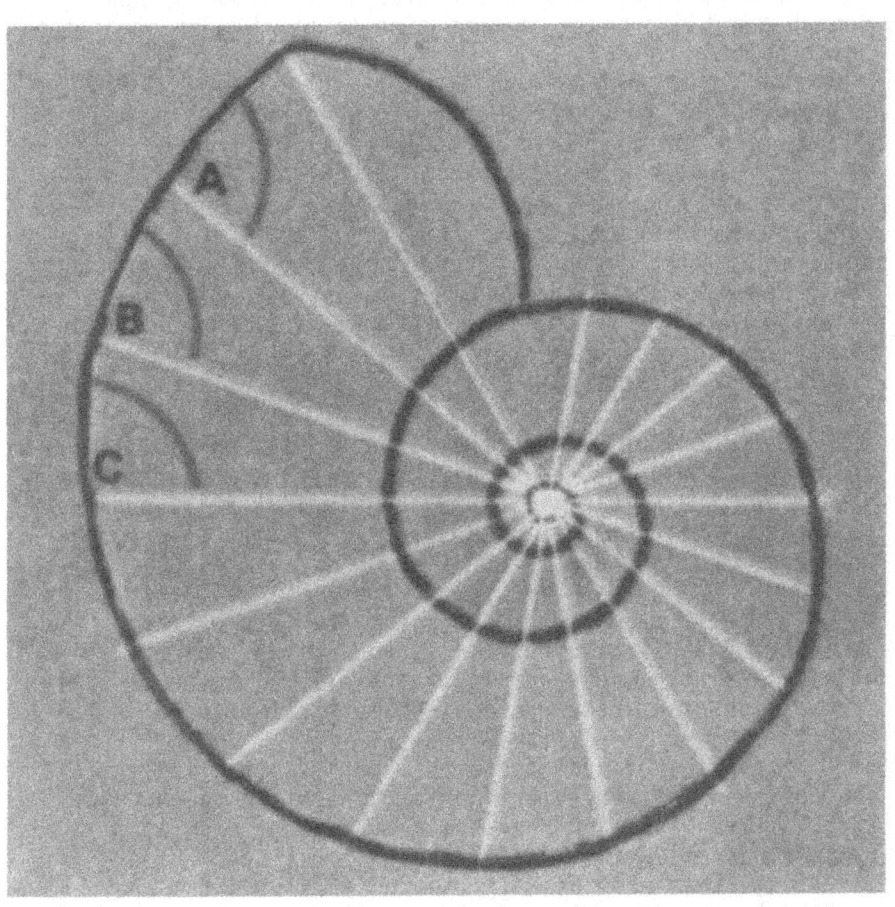

Figure 2
Source: "Mathematics", op.cit.
p. 93

The ancient Babylonians and Egyptians began to use the concept of measurement in formal ways dealing with the concept of length, area, and volume. One of the first persons to formulate the abstract idea of physical space was the Greek merchant, mathematician, and philosopher, Thales of Miletus (640-540 BC) "....Thales (for example) knew enough about astronomy at that time to be able to predict the solar eclipse of 585 BC".[1]

The Greek and Roman architects used precise mathematical proportions and geometric forms in developing their building designs. The height and location of columns in Greek temples, for example, were "generally in accordance with the...pythogorean triangles whose sides were in proportion to three: four: five".[2] Even the number of grooves and their depth in the Greek temple columns was based on a triangle with the proportions of three: four: five which the Greeks considered to be one of the most perfect geometrical figures. (See figs. 3 and 4)

The Greeks held "that the perfect forms of beauty" were the simplest geometrical figures, namely the triangle, the square and the circle. To them the simplest relations seemed to "determine the beauty of a shape"[3] (just as they did in the harmony of sounds). (See fig. 5)

In the 12th and 13th centuries, the work of the designers, architects, engineers, and builders of the magnificent gothic cathedrals, which exists in Europe today "reveals constant recurrence of geometrical proportions".[4]

Historians of architecture..."noticed that medieval churches had a fixed relationship between the height and width of the nave, and then they had fixed proportions in projection, elevation and all three dimensions."[5] In addition, these were used in the "decorative arts" and were applied in the designing of alters, fonts, etc. The Baldachins, for example, "to all appearances are of the wildest fantasy, yet they are composed of simply geometrical figures".[6]

It is interesting to note that the Gothic engineers/ architects used a form of applied descriptive geometry principles to design and lay out self-supporting arches that are critical to the structural integrity of the gothic cathedral. These principles included a form of orthographic projection and the concept of rotation. This can be seen in some of the actual graphic layouts shown in Figures 6 and 7. These are typical of layouts which were used to design and build the actual arches in gothic cathedrals.

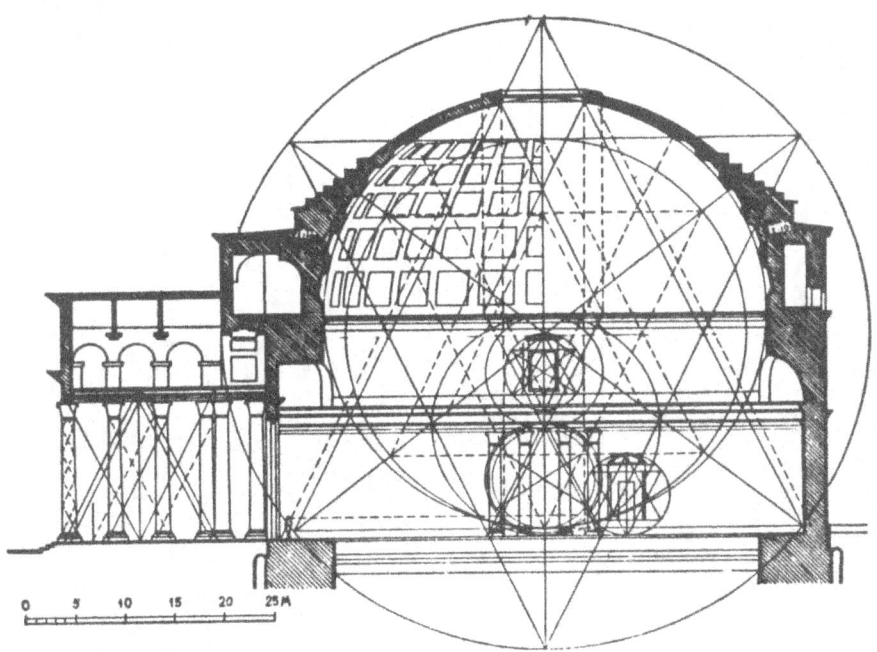

Figure 3
Rome's Pantheon

Source: "History of Aesthetics" Vol. I,
by Wladyslaw Tatarkiewicz, Polish
Scientific Pub. Warszawa, Poland,
1970, p. 56

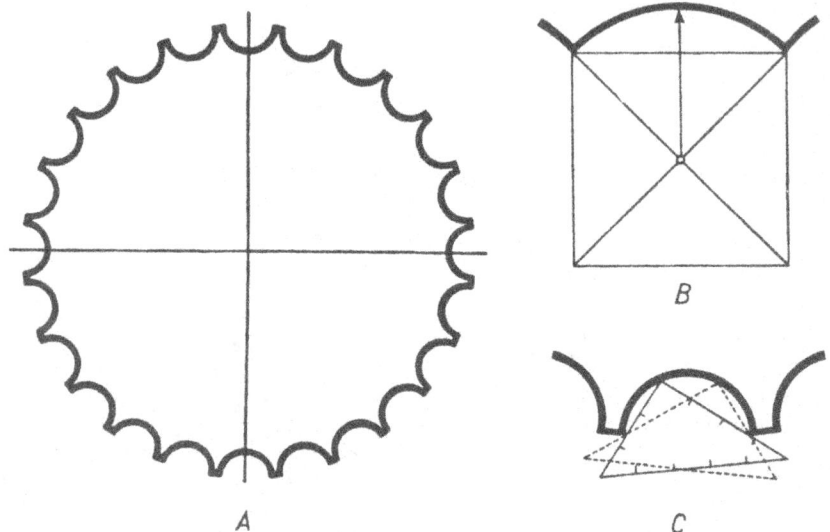

Figure 4
Grooves in Greek Columns

Source: Tatarkiewicz, op. cit.
p. 55

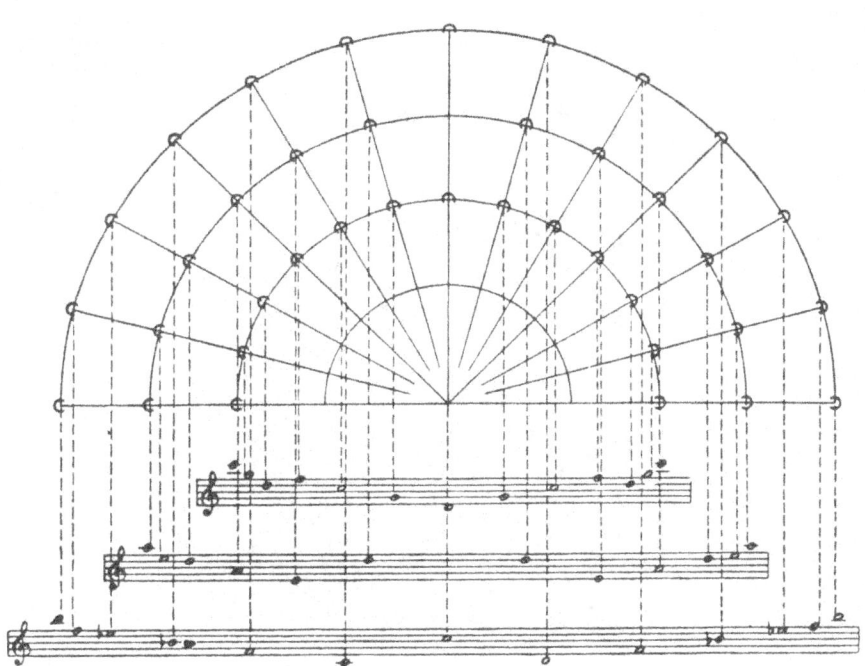

Figure 5

Source: Tatarkiewicz, op. cit.
p. 57

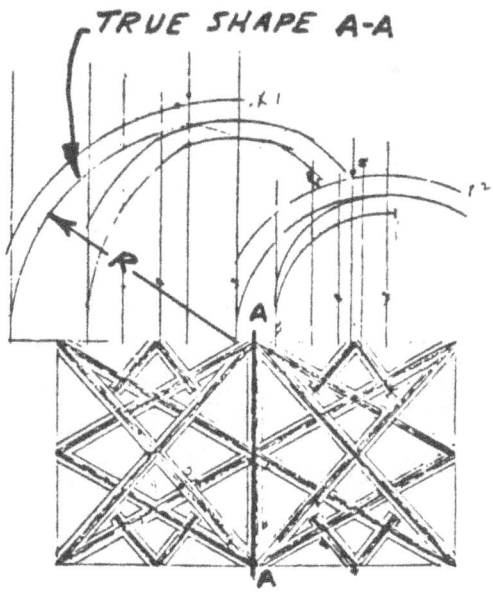

Figure 6

Source: "The Dresden Sketch Book
of Gothic Vault Projection" by
F. Bucher, Jan. 31, 1969 (unpublished paper)

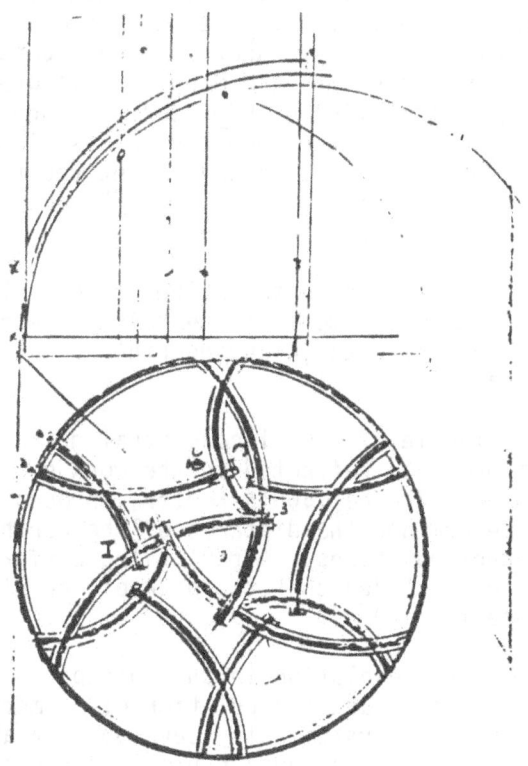

Figure 7

Source: F. Bucher, op. cit.

A close examination of Figure 6 for example, shows that the arch designated by A-A (notations added by SMS) in the horizontal projection is seen as a true shape in what is the equivalent of the true shape "diagram". This diagram is derived by rotating the plane of arch A-A into a frontal projection plane. (Rotation arcs are not shown since the arc of rotation was evidently "recorded" in the mind of the designer.) The method used here by the medieval architect-engineer reflects very closely the "direct method" used in descriptive geometry today. What is significant is that this approach pre-dates Gaspard Monge's "discovery" of descriptive geometry by about 500 years. Monge formalized his concepts of descriptive geometry in his "Geometries descriptive" published in 1795 and he set the tone for future development of this science through his "trace of a plane" spatial analysis approach.

Further evolution of descriptive geometry led to the formalization and refinement of the "direct method" which is largely the standard used in engineering education today in the United States.

Another example of visual conceptual graphics is in the field of astronomy as reflected in the work of Johannes Kepler. In the 1500's Kepler, through triangulation, graphically determined the distance of the earth from the sun at different positions. After trying different patterns for orbits, he concluded that the path of the planets was an ellipse. (See fig. 8)

Continuing the evolution in the conceptual graphical-visual representation and manipulation of geometry, H. Grassman in 1844, systemically developed an abstract geometry of several dimensions in his Austehingslehre (theory of expansion). Ludwig Eckhart in 1928 published his "Der vierdimensionale Raum" (Four-Dimensional Space), (translated by A. L. Bigelow and S. M. Slaby, Indiana University Press, 1968) in which he represented "Structures of Four-Dimensional Space by means of figures on the drawing board".[7] Figure 9 shows Eckhart's representation of a point in a 4-D space in which the techniques of the "direct method" of three-dimensional descriptive geometry is extended to the fourth dimension. Figure 10 shows the projects of a super-sphere (the four-dimensional analog of the sphere).

In 1965, Louisa Bonfiglioli (of the Technion in Israel) published a number of papers dealing with multi-dimensional space while she was with the Department of Graphics and Engineering Drawing at Princeton University including one entitled "Parallel Projection for Euclidean Geometry of

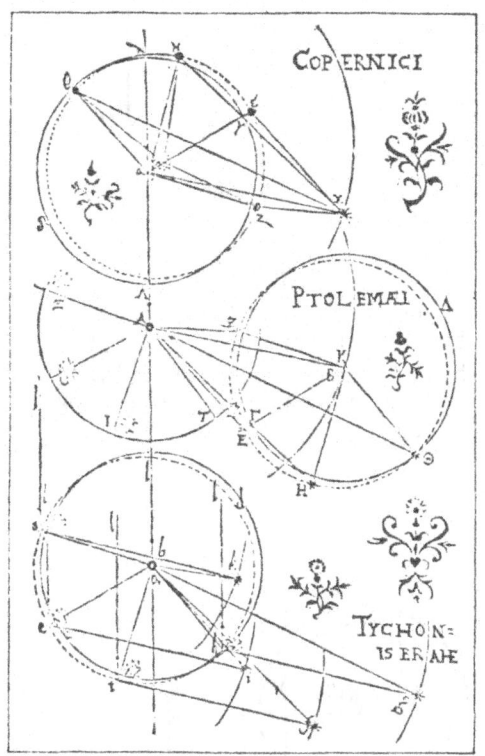

Figure 8

Source: New York Times,
Feb. 19, 1972

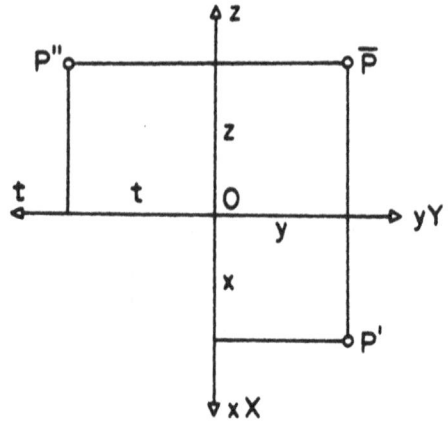

Figure 9

Source: "Four-Dimensional Space"
by Ludvig Eckhart (translation
by A.L. Bigelow & S.M. Slaby)
Indiana University Press, 1968,
p. 40

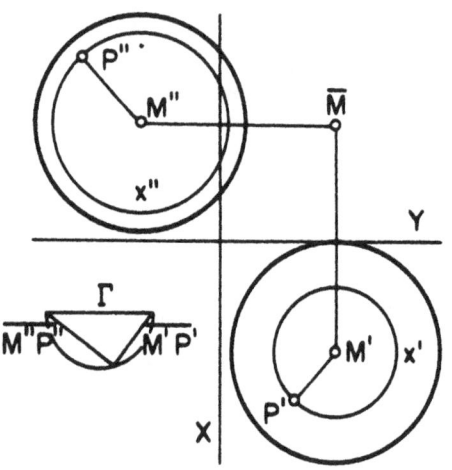

Figure 10

Source: Eckhart, op.cit., p. 82

Four-Dimensional Descriptive Geometry" (McGraw-Hill, Inc., 1968), where the "trace of a plane" and "direct" method are both used to represent and manipulate 4-D spatial concepts. (See fig. 11)

In Forest Woodworth's "Graphical Simulation" published by the International Textbook Company (1967), he shows how an, extension of the direct method of three-dimensional descriptive geometry is applied to draw an edge view of a tetrahedron located in a four-dimensional space. The tetrahedron is represented as four non-coplanar lines determined by 4 points located in a four-dimensional system. (See fig. 12)

Applications of four-dimensional descriptive geometry concepts are apparent in some of the attempts made to visualize Einstein's space-time continuum and in areas of electrical engineering and computer science, as well as in the field of statistics which plays such an important role in physics and engineering.

The limits to the conceptual development of visual/ graphical techniques to reveal geometrical/spatial relation- ships have not been approached. More theoretical development is possible and inevitable and this will be illustrated by the papers that follow. It is significant to remark that some of the most interesting work in multi-dimensional geometry or "hypergraphics" is being done by artists, such as David and Harriet Brisson at the Rhode Island School of Design and Dennis Finch, a free lance artist.

The connection between science and art was clearly stated by Aristotle, in his <u>Analytica</u> where he wrote "An Experience, that is, the Universal, when established as a whole in the soul - the One that corresponds to the Many, the unity that is identically present in them all - provides the starting-point of art and science: art in the world of process and science in the world of facts".[8]

The combination of art and science has led to unprecedented technological developments, even though the connection between the two has not always been openly admitted. Geometry is the "glue" that binds together, art, science, technology, and life!

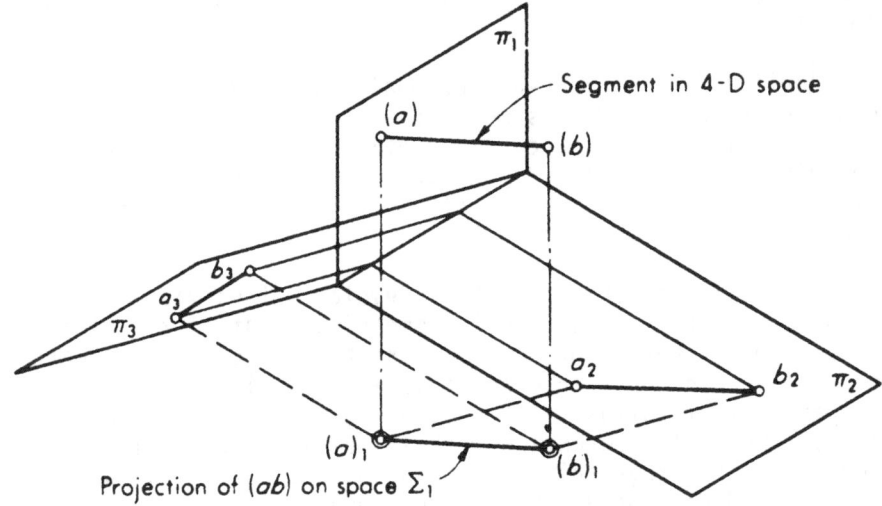

Figure 11

Line Parallel to a Space of Projection. From <u>Four-Dimensional Descriptive Geometry</u>, by C.E.S. Lindgren & S.M. Slaby. Copyright©1968 by McGraw-Hill Book Company. Used with permission of McGraw-Hill Book Company.

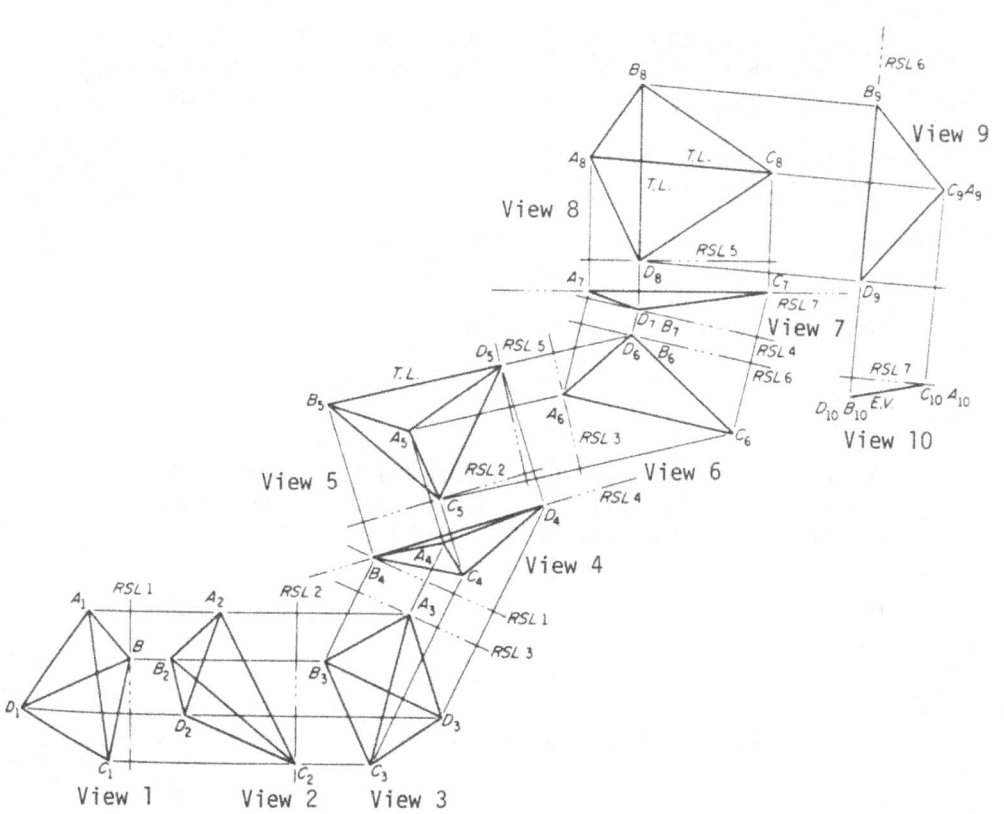

Figure 12

Source: "Graphical Simulation" by
Forest Woodworth, International
Textbook Co., Scranton, Pa., 1967,
p. 487

FOOTNOTES

[1] Adler, Irving, "A New Look at Geometry", The John Day Co., N.Y., 1966, p. 15.

[2] Tatarkiewicz, Wladyslav, "History of Aesthetics", Vol. 1, Polish Scientific Publishers, Warszawa, Poland, 1970, p. 54.

[3] Ibid. p. 51.

[4] Ibid, p. 153.

[5] Tatarkiewicz, Vol. II, 1970, p. 161.

[6] Ibid.

[7] Eckhart, Ludwig, "Four-Dimensional Space" (Translation by A. L. Bigelow and S. M. Slaby) Indiana University Press, Bloomington, Indiana, 1968, p. 39.

[8] Tatarkiewicz, Vol. I, p. 156.

REFERENCES

Adler, Irving, "A New Look at Geometry", The John Day Co., N.Y., 1966

Bergamini, David, "Mathematics", Life Science Library, Time Inc., N.Y., 1963

Eckhart, Ludvig, "Four-Dimensional Space" (translation by A. L. Bigelow & S. M. Slaby), Indiana University Press, Bloomington, Indiana, 1968

Lindgren, C.E.S. & Slaby, S.M., "Four-Dimensional Descriptive Geometry", McGraw-Hill Book Co. Inc., 1968

New York Times, February 18, 1972

Tatarkiewicz, Wladyslav, "History of Aesthetics", Vols. I & II, Scientific Publishers, Warszawa, Poland, 1970

Woodworth, Forrest, "Graphical Simulation", International Textbook Co., Scranton, Pa., 1967

2

Speculations on Dimensionality, Valence and Aggregation

Cyril Stanley Smith

Abstract. Using a pictorial-intuitive approach, dimensionality is regarded as an aspect of connectivity between vertices of various valencies. Existence without interaction is ignored. The valence of a vertex ("thing") arises out of its substructure and is dependent on the scale of resolution of the means of interaction and observation. A two-dimensional projection of connected vertices preserves (with some ambiguity due to superposition) the connectivity of a system of any dimensionality. The world seems to be three dimensional because in three or more dimensions any two vertices can be directly interconnected regardless of intervening structure.

All physical systems involve tension and compression in balance. Change occurs by catastrophic buckling at some level of structure but is unnoticed at others both above and below. The formation of a kind of moiré pattern from misfit at the interface between two nearly but not quite identical structures may provide nuclei for new ones.

A work of art can be considered as a multivalent vertex with internal structure, capable of suggesting in the mind of the viewer relations between previously unrelated things, both internal and external, and of generating entirely new insight from the new structures that are formed by the slight misfits that result.

Introduction

The viewpoint adopted in this paper is more speculative than rigorously theoretical. Elsewhere (1) I have discussed the role of art, or at least aesthetic curiosity, in the discovery of technologically useful materials and techniques. At another scale of observation, I have suggested (2) that the separation of phases in an alloy or other chemical system

Figure 1

involves the same relationship between unit and immediate environment, of region and overall connectivity, as that which gives interest to a painting, a landscape or a viable society. None of these can be understood analytically; all depend on different levels of hierarchical interaction. My aim is to discuss some aspects of connection theory in an elementary way and to suggest that in complex systems resonant communication is essentially one dimensional when viewed at the pertinent scale of observation. This permits the representation in two dimensional form of relations in multidimensional systems, yielding visualizable patterns in which the overtones of meaning are more easily experienced and understood than when the same statements are given in mathematical notation or verbal forms of logic. This paper is intentionally somewhat naïve in the hope that it may contribute in a small way to the resolution of seeming conflicts between art and science, between thing and environment, and between individual and society. The reader however must be aware that it deals with the establishment of connections and not, as in rigorous dimensional topology, with the exclusion of connections by closure. Some measure of choice, of history, is allowed. Though the diagrams resemble simplistic mechanical drawings, they are not intended to imply any mechanism but merely to map the connections between nodes in interpenetrating and partly interacting resonances. Though definite in a topological sense they are deformable and must not be thought to represent reality, whatever that is.

On Things and Environments

Everything, every "thing", takes meaning only by interaction with something else. Indeed its existence cannot be known without such interaction. The inside of Fig. 1 represents this, though its boundary is joined to the rest of the world. Both physically and conceptually the qualities of things are discernable only by their response to an external probe interacting with their internal structure. The real units of the world are not particles - atoms, in whatever sense one uses the word - but connections, and a connection must be between two things. This duality is the basic symmetry, the only necessary one, but if the "things" have internal structure they may be more than monovalent and provide nodes for combinations of connections with other connections to give higher symmetries. If complex units can interact as wholes with each other, the connection itself will be complex but it is still one-to-one in relevance. Though it is a complex prism, it serves dimensionally as a line joining matching patterns within the things at each end. However, within a "thing" of complex inner structure the parts may react differently to different types of communication, so resulting in different overall valences: Different parts may either belong to entirely separate networks (interpenetrating but not interacting) or they may share certain nodal points and interlock as in the case of magnetic interaction between atoms in a crystal lattice or an individual as member of both a family and an intellectual community. It is all a question of what connections are actually established at the time of observation. Moreover, the "thing", unlike the mathematician's ideal zero-dimensional point, is physically a complex network of internal structure, itself forever hierarchical. A "thing" is seen as such only on a scale which has been chosen to allow the internal valencies to have cancelled, leaving visible only connections with other things on its own level, with perhaps some connections extending entirely out of the field of view.

The emphasis on simple connectivity excludes those topological features that are dependent on the genus of a system. Both a torus and a sphere in the absence of external connections are treated as zero dimensional, while internally they are three and four dimensional, for the sphere requires three trivalent vertices and the torus four quadrivalent vertices to define it. A melon with $\underline{n}$ lobes joining at each of two vertices is to be regarded, connectively, as the simplest $\underline{n}$-dimensional object.

Polyhedra don't exist

There has been much speculation as to why the world in which we live seems to be three dimensional. Our experience as children learning by moving and poking and acquiring a sense of inside, outside, flexibility and solidity makes us think of dimensionality in terms of points, polygons and polyhedra, the things so beautifully related by Euler's Law. I contend that polygons and polyhedra do not exist but only what Arthur Loeb in private conversation has called polyvertices. Polygons and polyhedra mark the negation of connections, and their seeming reality is a mere construct of the human senses responding to gradients of density of vertices with one-dimensional interconnections. Both polygons and polyhedra can be shrunk to points without any change in external connections. Whether or not they appear is entirely a matter of choice, of the scale or resolution of the means of observation.

The Dutch artist Maurits Escher, who has enriched the understanding of many subtle points of geometry and topology, illustrated this in his lithograph <u>Kubische Ruimteverdeling</u> (3). Cubic space filling is not the packing of cubes face to

face, edge to edge and vertex to vertex, but is simply a
connected network of linear connections between hexavalent
nodes. In any system the simple relation

$$\sum rV_r = 2E \quad \ldots \quad 1)$$

must hold, where V_r is a number of vertices of valence r and
E is the number of connections between them. (Connections
severed for convenience of local study are assigned vertices
of valence one). This involves no consideration of dimensionality beyond the basic duality of connection. Even complicated systems reduce to the closure between just two vertices, each of which has valence (for the connections under
consideration) equal to what is usually called the dimensionality. As larger and larger regions are examined in an
extended system it may either aggregate or cancel.

Dimensionality

Dimensionality in the limited sense used here can be
determined by examining the connectivity of vertices forming
the network of closure that defines a "thing". Although the
closure may surround internal structure and although it may
itself be embedded in a larger system of any dimensionality,
its own dimensionality depends upon the number of different
routes of connection between any two of the vertices on the
boundary that have the highest valence for internal connections. It is a property of the vertices' intercommunication.
Fig. 2 shows the simplest cases. An isolated point has
valence and dimensionality both zero. In Fig. 2A, a one-
dimensional system, there is a single connection between the
two monovalent vertices. Fig. 2B and 2C have two and three
dimensions respectively, corresponding to the valencies of
the two vertices. Any of these figures may form a component
embedded in a system of higher dimensionality, but there are
only r routes emanating from and terminating at each of the
two vertices situated on the boundary and these serve to define its dimensionality for our purpose.

Note that the valencies are actual connections satisfied within the system selected for study, ignoring any external connections or potential ones, and the dimensionality
does not depend on any supposed quality of space. Note also
that the recognition of a connection is already a question
of hierarchy and scale, for an externally operable valence
of anything other than an ideal mathematical point is a consequence of an internally unsatisfied valence; one not satisfied within whatever has been chosen to define closure, and
this in turn will affect the dimensionality of anything in-
corporating it. Figure 3A is a representation of either a

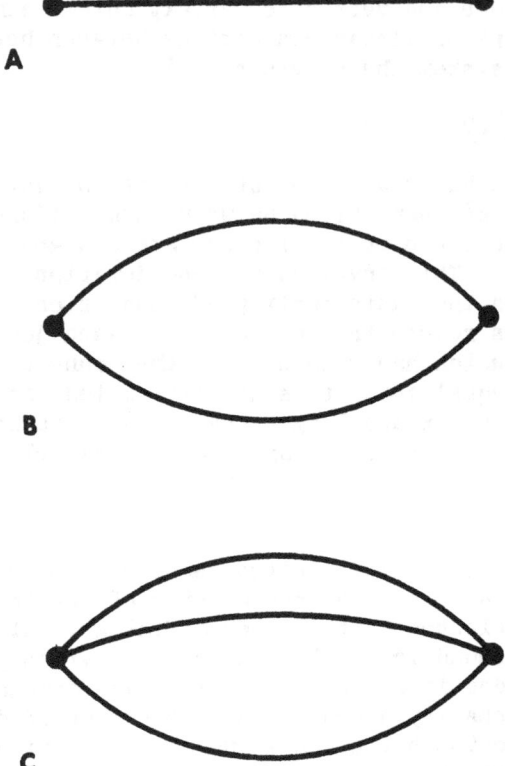

Figure 2

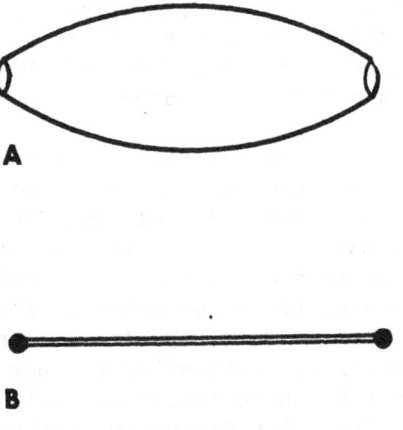

Figure 3

two dimensional object or a three dimensional one depending
on whether or not one chooses to see the small structures
within the vertices at the ends, and Figure 3B is one- or
two-dimensional depending on whether the connection between
the vertices is resolved into two lines or is seen as one.
Dimensionality becomes a function of what can be separately
treated as communications. Mathematically, dimensions are
distinguishable; in the physical world they can merge or separate or disappear entirely depending on the resolving power
of the means of observation and interpretation - the "dis-
cernability" of Leibniz, so well discussed by Morrison (4).
Opposing but equal symmetries cancel each other out after a
few replications within a uniform aggregate; the departures
from symmetry may be cumulative and eventually come to domi-
nate larger relationships until eventually they too become
submerged.

Moreover, although in two dimensions closure defines inside and outside by denying access to a region without cutting the boundary, in what we call three dimensions there is
no such restriction, and everything is both inside and outside: communication can bypass any obstacle and directly
reach any vertex anywhere which has an available valence.
This applies also, without limit or difference, to any number
of dimensions above two.

Since the linear connections between vertices are all
that matter, the whole hierarchy of connectivity within a
structure of any dimensionality is retained on projection to
2 dimensions. The projection of the vertices and edges of any
isolated polyhedron is a polygon with a closed one-dimensional boundary that includes within it vertices in number and
valence mapping all of the original ones. Except in projections from points within the polyhedron (or in its surfaces
as in a Schlegel projection) there will be additional pseudo-
vertices originating in the superposition of projections of
edges on opposite sides of the original. They are a consequence of the simplified definition of dimensionality we have
used, but they are all quadrivalent and their presence does
not change the connectivity of the original vertices.

Dimensionality is a statement of the relation of things
to their environments; it is basically hierarchical. Everything, however complicated internally, is zero-dimensional
unless it has external connections. Dimensionality is a
property only of connections on and within some defining
boundary. In an extended array, the dimensionality of the
connections between any two points on the closed boundary of
a region selected for study will oscillate as the boundary is
moved out from an initial vertex and connections internally

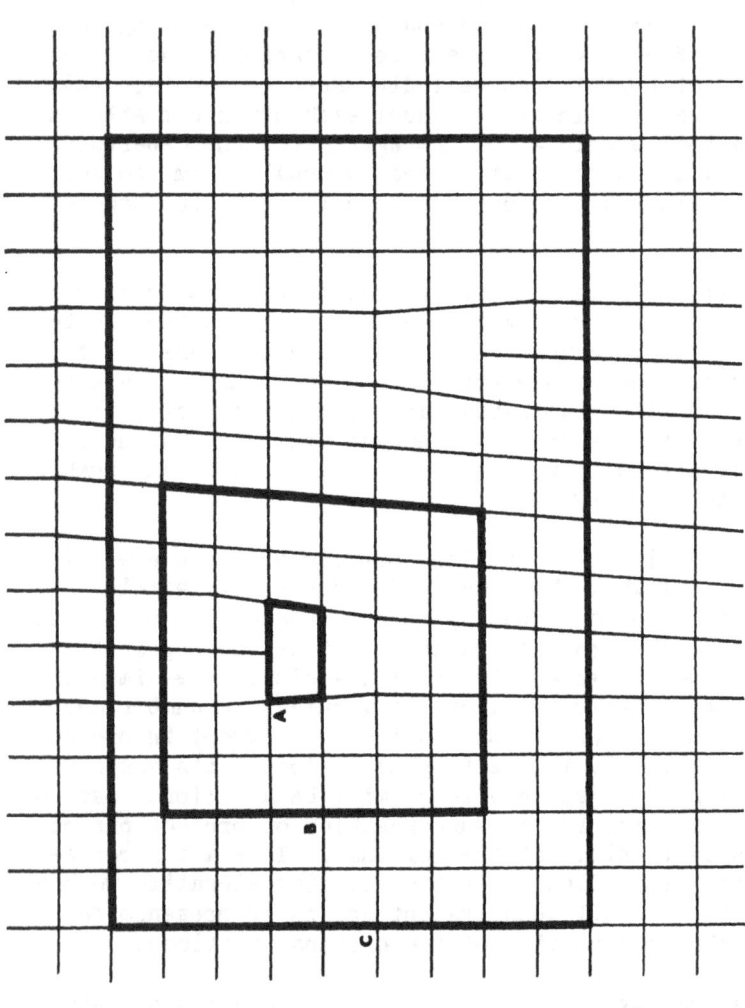

Figure 4. Quadrilateral lattice of quadrivalent vertices, with imperfections. The connections extending outward and inward from the vertices on the emphasized boundaries of the subgroups have the following relations: For boundary ($\underline{A}$), $E_o - E_i = 9 - 0 = 0$; ($\underline{B}$), $27 - 19 = 8$; ($\underline{C}$), $50 - 42 = 8$. See equations 2 and 3.

diverge or join and compensate each other.

If, as in a crystal lattice, a network is formed by the iteration of a space-filling unit cell with internal structure the two-dimensional projection of the assembly will be divisible into tessellations that also are space-filling by iteration. What is left over after internal satisfaction of valencies is available for establishing external connections. An examination of only the vertices on a given boundary will show whether or not it defines a region that is iteratively space-filling, and of what dimensionality. See Fig. 4. Whether in two dimensions or the projection of a multi-dimensional structure into two, the closure of a possible unit cell is by a boundary from the vertices on which the outreaching connections outnumber the inreaching ones by the following relations

$$E_o + E_i = V_b(\overline{r}_b - 2) \quad . \quad . \quad . \quad . \quad . \quad 2)$$

where E_o and E_i are, respectively, the number of external and internal connections, V_b is the number of vertices on the boundary and $\overline{r}_b$ their average valence. (The closure of the boundary itself employs two valencies of each vertex). Also

$$E_o - E_i = \sum_i n_i(r_i - 2) \quad . \quad . \quad . \quad 3)$$

where n_i is the number of vertices on the boundary having valence r_i. If all vertices are of the same valence, without regard to the number of sides on the polygons formed between them, then if

$$E_o - E_i = 2r \quad . \quad . \quad . \quad . \quad . \quad 4)$$

the average polygon within the boundary has $2r/(r - 2)$ sides and the array could be replicated to fill space indefinitely. If a boundary is moved through a simple space-filling array of polygons equation 4 will be satisfied whatever its location, but even if the array is not space- filling the value of $E_o - E_i$ will remain constant as long as the boundary is in a region that is on the same hierarchical level, for except at the boundary of a central anomaly each additional E_o is accompanied by an E_i.

When vertices are not all of the same valence and equation 4 is not immediately satisfied, it is always possible to outline groups within contiguous nets of quadrilaterals or hexagons superimposed upon the actual vertices but disregarding some of their valencies. These correspond to the well-known square or hexagonal tiling or their duals. Each vertex has its opposite and is shared by r-1 contiguous cells.

A treatment of the compatability of group tessellations based on internal polygons and their vertices is given by A. L. Loeb in his excellent Space Structures (5). However, the space-filling quality is a matter of a boundary valencies only and is not affected by any internal complication. The E_o - E_i count on a sample excised at any place from an array will tell immediately whether the group could be replicated to form an extended aggregate. It has the useful property that, like the related metric Burgers' vector of the crystallographer, it will immediately reveal the presence of uncompensated dislocations within any region of a lattice (Fig. 4). If it remains constant as the boundary is moved, the boundary has remained in the same region of superstructure. A change in its value signals vertices lying on the interface between two units in a higher level of hierarchical superstructure. Only if equation 3 is satisfied and if both E_o and E_i are even can the array be space filling. It should be noted that the external connections from any region do not derive from averages of what is inside but only from exact sums of connections. Statistical analysis would submerge the points of interest, for the aspect of an array that determines mere extension or provides the critical step to a higher hierarchical level may reside in the ability of a single individual among thousands to make a new type of connection.

Dimensionality from interlock

The generation of a three dimensional structure by the intersection of three one dimensional ones is shown in projection in Figure 5. Parts of this are one dimensional, two dimensional or three dimensional depending entirely on what vertices are included in the sample chosen for study. Connecting only vertices A and B represents a one dimensional assembly; A, B and C, two dimensional, while including D makes it three. The heavily outlined area can be seen as a two dimensional group of 3 rhombi forming a hexagon that could join with three polygons like itself (each having substructure) at each vertex to form an extended two dimensional array (Fig. 6) or it could be regarded as the projection of half a cube closed with a doubly-curved hexagon. It is already connectively three dimensional, but it becomes a conventional cube if the hidden edges are added to make all vertices internally trivalent. A shift in the direction of projection to avoid superposition of the inner vertices gives the common perspective view of a cube, Figure 6A, still within a hexagonal outline. If there were no external connections, i.e. if all valencies were satisfied within the selected boundary, the set would then behave as a zero-dimensional inclusion within any larger grouping, but if periph-

eral valencies allow outreaching connections it may join adjacent hexagons to become a part of an extended aggregate
either two or three dimensional. With all vertices hexavalent (six peripipheral and two internal ones when viewed
two-dimensionally) it is a projection of a simple cubic lattice. The unit cells of all crystal classes are projectable
into a hexagon with internal and external matching of vertex
valencies, and the many levels of crystallographic superstructure are representable by internal connections within
larger hexagonal groupings. This underlies the elegant
treatment of A. F. Wells (6) as well as the cosmic overtones
that Moslems see in their mosaic patterns.

Note how the nesting of connections within a 2-D network
closure has generated a 3-D closure. The extensions from
the trivalent vertices of the projected cube in the center
of Figure 6 have increased the valencies of the vertices on
the hexagonal boundary and hence the dimensionality of what it
defines. Just the same occurs with higher dimensions. The
hexagon within a hexagon of Fig. 7 is a projection of a cube
joined within a cube: it is a hypercube with all vertices
quadrivalent. It could join with others like itself to give
an aggregate in the form of a four-dimensional lattice of
cubes meeting at 14-valent vertices, somewhat as suggested
in Fig. 8 in which the diagonal connections are omitted except in the center to avoid confusion. It could also be
joined, not as a lattice but by extension to meet an external closed shell of pentavalent vertices - a five-dimensional hypercube which can still be projected as a hexagon or
space-filling polyhedron (Fig. 9). This characteristic of
nesting or condensation of connections within shells, is the
very basis of hierarchy and it can be represented in two dimensions. If all valences are satisfied internally, i.e. if
there are no outreaching connections, the object when viewed
externally is effectively zero dimensional, a single "thing",
but peripheral valences may join it to other things like itself (either individually or in groups) to form a continuing
aggregate.

Crystals

The aggregation of atoms into crystals with internal imperfections provides a good example of these relationships.
It is interesting to see the different way in which crystals
have been regarded by scientsts at different times in history,
for the history of science lies largely in the sequence of
discovery of significant hierarchical levels of structure
(and temporary overemphasis on the newly discovered ones).
The first view of a crystal was as a polyhedron, of interest
only for its more or less regular exterior geometry. Then

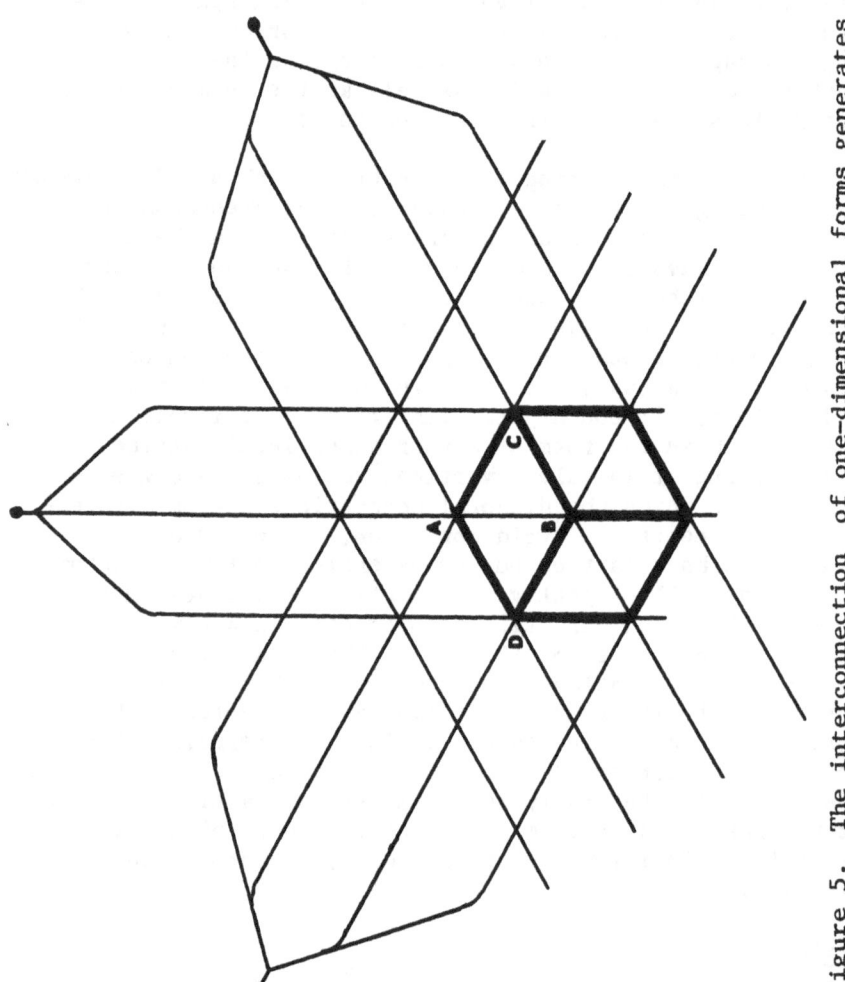

Figure 5. The interconnection of one-dimensional forms generates two- and three-dimensional ones.

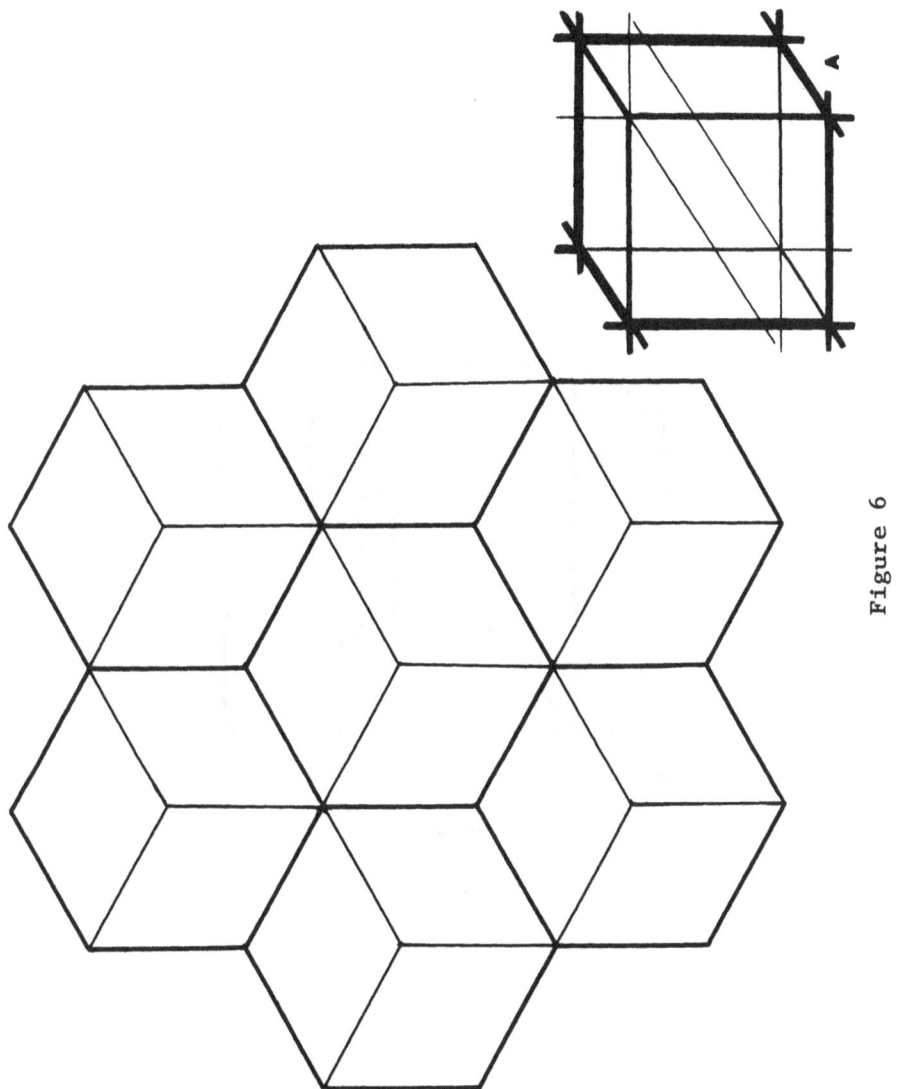

Figure 6

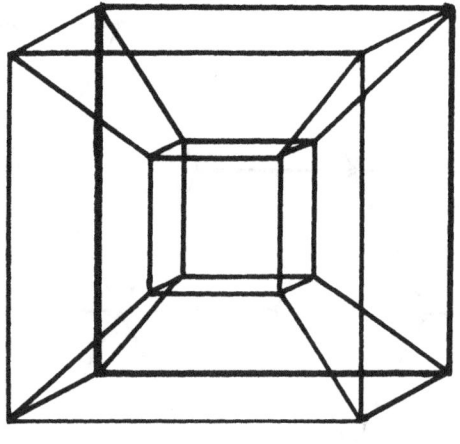

Figure 7

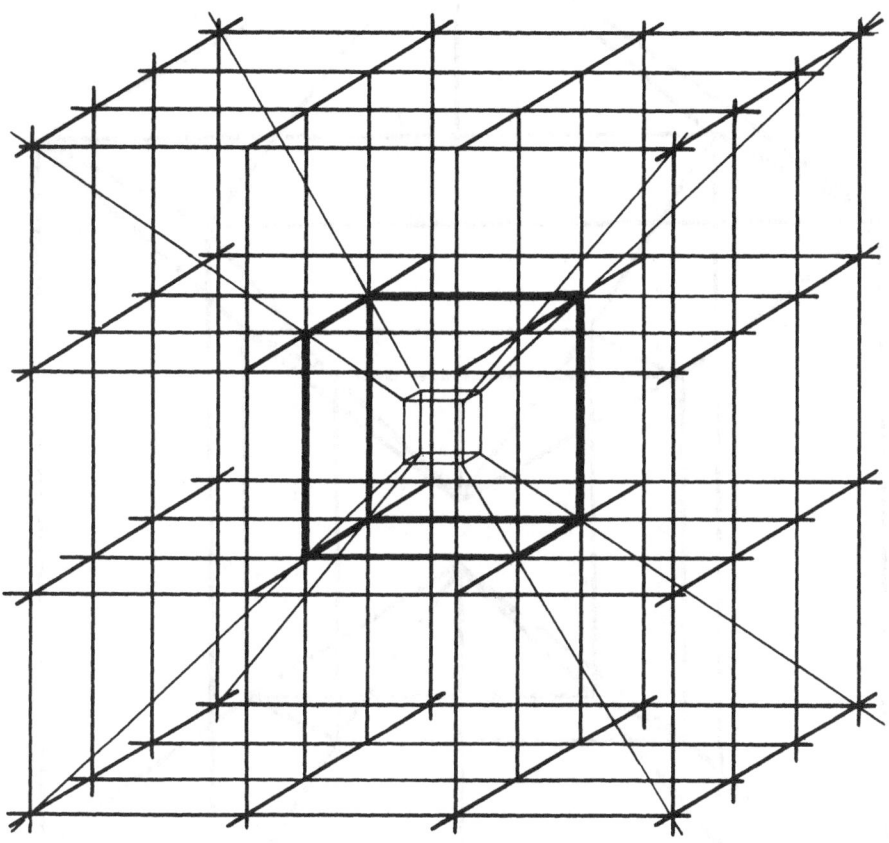

Figure 8

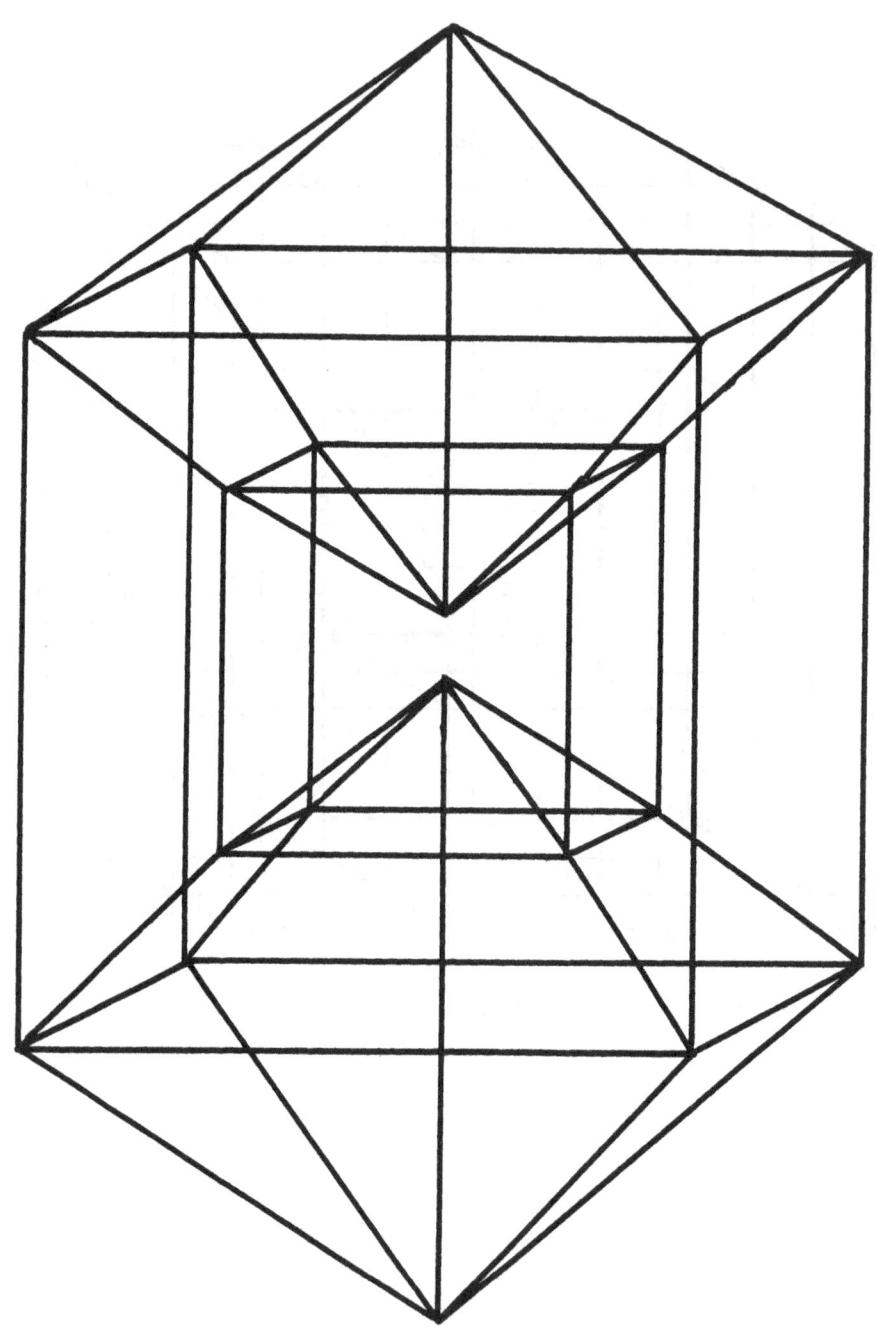

Figure 9

came Kepler's idea (long known and used in the decorative arts) that geometric symmetry could emerge from the stacking of isotropic parts. With the substitution of ball-like atoms for the invisible drops of mist in Kepler's snowflakes this is essentially the freshman's view of crystals today. This model was made good use of by Hooke and others in the 17th century but curiously it was replaced by stacked polyhedra, essentially uniform cleavage fragments re-assembled face to face to fill space when later scientists came to value respectable mathematical models over intuitive physical ones. After Dalton, the Keplerian model could return, but it was not generally accepted until the very end of the 19th century (7). Then it had no sooner been beautifully confirmed by x-ray diffraction than it was exploded into probabilistic clouds of electrons visualizable at best in Patterson plots of their averaged densities!

Two common drawings of crystals in elementary texts today are as in Figures 10A and B, which emphasize either the unit atoms in contact or their connections within the regular lattice, here 2-dimensional for simplicity. Better is Figure 10C, showing both the atoms and the connections, but the most meaningful might be a diagram such as Figure 11 based on the unit shown in Figure 12 wherein the internal structure of the atoms is shown in relation to the exterior, with an interlocking hierarchy of bonds between nodal electrons, the whole being geometrically distortable but not topologically. Figure 11 illustrates several stages in this kind of connection and suggests how they could continue downward into individuality or upward into a lattice almost without limit. For study, we select some part of a hierarchy by cutting it out of its environment. There are centers and interfaces, alternations of high and low density of connections; there is a closure of connections within a given clump of nodes and usually some left over for external interaction. Figure 12 shows that the external quadrivalency of the atoms in Figure 10 is based on the exclusively trivalent internal vertices transmitting outward the incongruities of the central group of four with an ever-decreasing density. Had the core been a hexagon with the same trivalent vertices the connections would have been satisfied within a uniform lattice of hexagons with no centered anomaly or density gradient. The same would follow -- as it does on the larger level of Figures 10B and 10C -- with quadrivalent vertices in quadrilateral array.

As discussed above, a count of the valencies extending beyond any closed boundary can tell whether the enclosed region contains a singularity or if it could be a part of a uniformly extending region within which differences have been

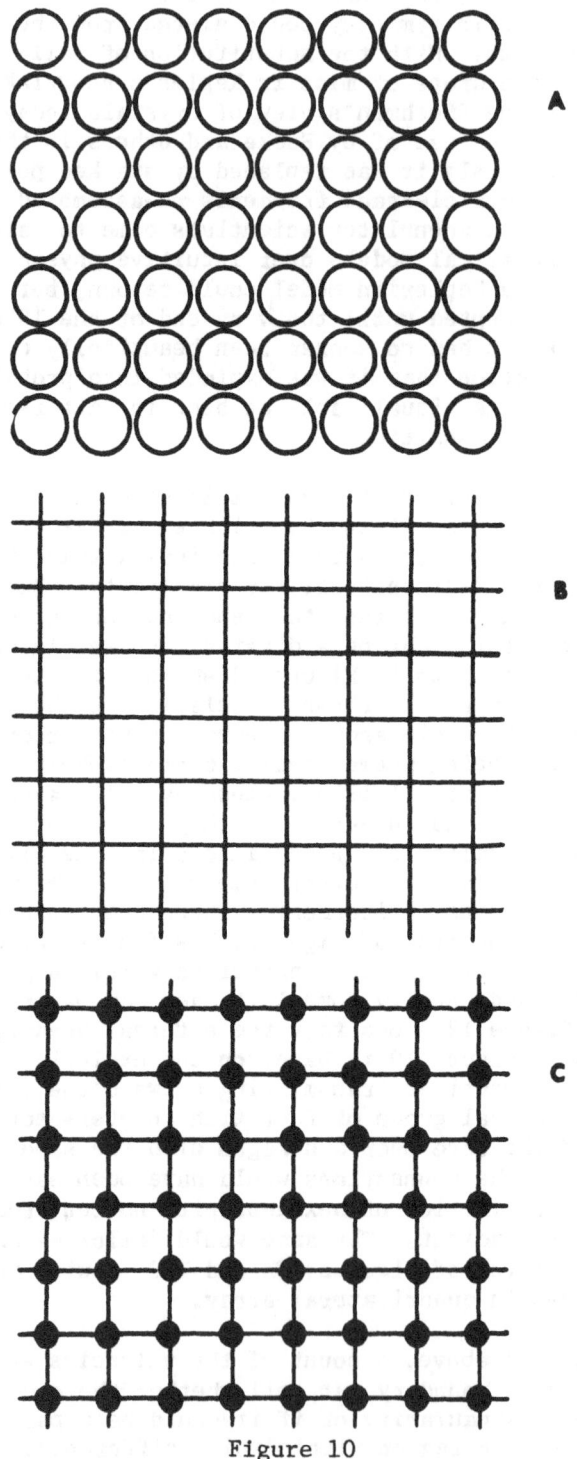

Figure 10

Figure 11

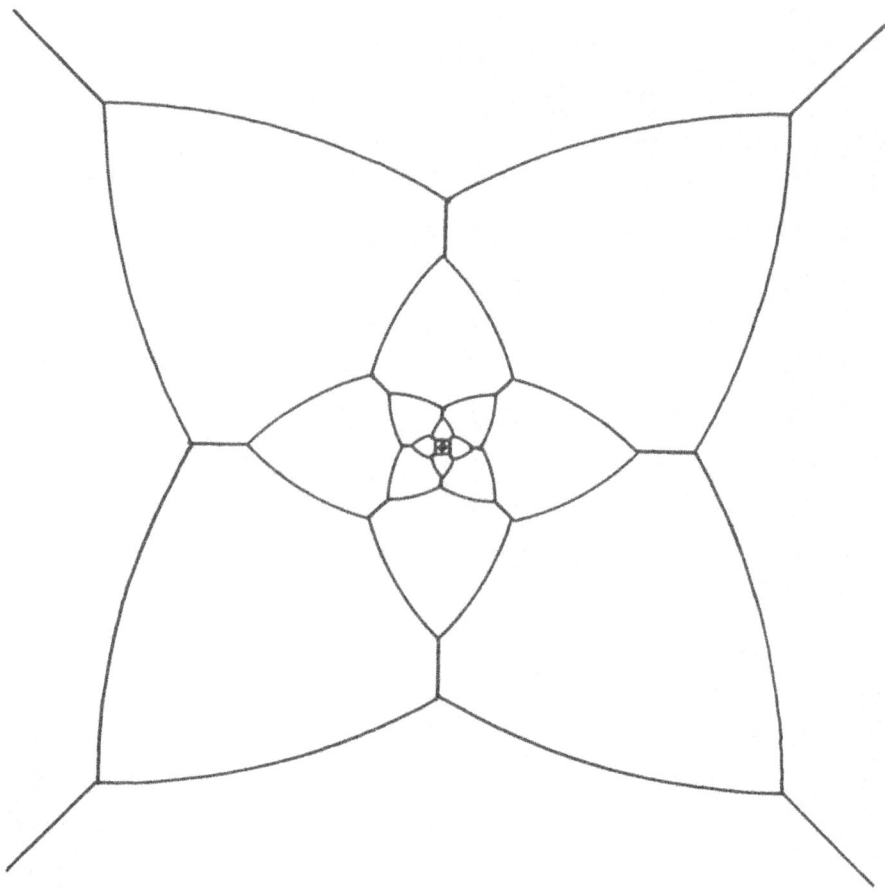

Figure 12

locally cancelled. Mechanical imperfections in crystal lattices were shown in Figure 4. They appear again in Figure 13
along with other types of imperfections arising from the
presence of foreign atoms in random, ordered, or segregated
locations. Any such local feature remains to characterize a
larger region of the lattice until its effects are cancelled
or lost at some larger boundary.

Perfect crystals are incredibly rare, and most crystal-
line materials are aggregates with internal boundaries mark-
ing first or second order discontinuities in composition or
orientation or longer range effects of balanced strain,
electrical or magnetic interaction. Within a region, however large, in which the relationship between neighbours remains unchanged, the region is properly regarded as homoge-
neous, when viewed at a lower scale of resolution.

On Stability

The closure of a boundary forming a feature of any dimensionality separates inside "from outside". With vertices
that are more than two valent there will be a difference between the density of connections on the two sides of the
boundary. Though mathematically there is no restriction,
any physical structure will collapse or expand without limit
unless both tension and compression are able to balance within a stable configuration. The necessity for triangulation
underlies the granular structure that appears in nature at
all levels. To resist buckling under compressive force there
must be a hierarchy of internal compression and external tension (or vice-versa) in balance within every closed unit in
isolation or within any cell in an aggregate. In three dimensions or less, buckling will occur unless there is restraint in $D + 1$ directions. Four restraints in three dimensions are the basis of solidity, but the addition of
higher dimensional connection does not necessitate more restraints.

Local buckling is the basic mechanism of any change
that is not disruption. The basic difference between the
states of matter - solid, liquid and gas - are the number of
restraints to the movement of the unit atoms or molecules
over each other. In a gas there is a single compressive restraint acting at the moment of collision; in liquids there
are three, which (in the three dimensional world) allows
particles to slide and to exchange neighbours but not to
separate; while in solids each particle is restrained by at
least four others. However, restraint on one scale is not
necessarily transferred to others. The resonant movement of
electrons and photons underlies the stability of 3-D struc-

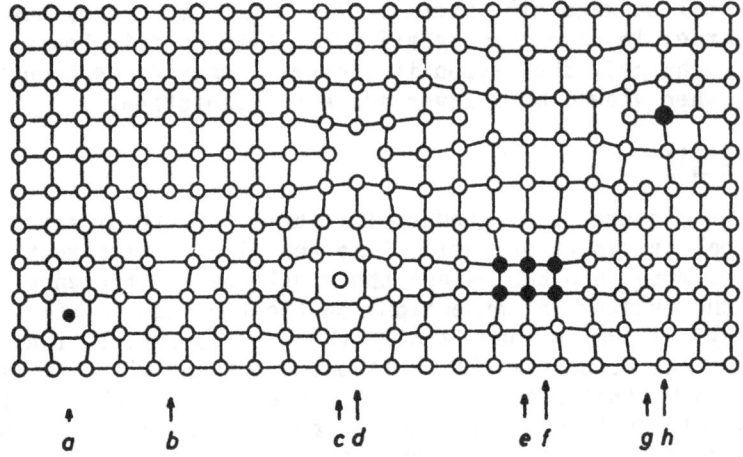

Figure 13
Imperfections in a square lattice. There are dislocations,
(b and g), lattice vacancies (d), and interstitial atoms (c).
Foreign atoms substitute on main lattice sites (h), appear
interstitially (a) or form segregated areas (e) -- all
without affecting the valence or near-neighbour relationship
at the periphery. (From H. Föll and B. Kolbesen, <u>Jahrbuch
der Akademie der Wissenschaften</u>, Göttingen, 1976.)

tures and stable triangulation (or better tetrahedroning, to use Buckminster Fuller's nice word) becomes unstable when the sizes of the components are grossly unequal, for a force has no effect in a direction normal to itself. So solid grains can form a component of a liquid or gas when fluidized, and, conversely, a mixture of two immiscible fluids can form a solid foam when droplets of one are surrounded by a minor volume of the other.

All change reduces at some level to the collapse of an arch under compression. The nucleation of a new structure within a previously stable array involves some local novelty of connection, the lateral removal of a stable tetrahedral block. Physically this is ultimately the insertion or removal of one atom between two others, the transfer of a bond from one neighbour to another. If the system is stable any small fluctuation is pulled or pushed back into conformity with the environment, but if some external change has made the structure metastable and if the fluctuation involves a sufficient number of new interactions to overcome the cooperative resistance of the old then the interface between the two structures will continually buckle locally. The zone of collapse will move laterally and in so doing will continually free the units to be rearranged into a structure better suited to the new conditions. The formation of the interface is the most difficult step, and it is most likely to occur in the vicinity of points of misfit between contiguous regions of complex structure in a preexisting hierarchy. If observed on a scale such that the units cannot be resolved this becomes the balance between volumetric and surface free energies in classical nucleation theory.

Some such mechanism seems to underly human perception, the communication and conception of ideas, and the reaction to a work of art: All of these depend upon the matching of preexisting patterns with new ones, or the making of new comparisons between previously unconnected parts of old ones. The mind can produce purely mental superstructures whether or not the initial stimulation involved physical connection.

The emergence of new patterns of different dimension and dimensionality from the superposition of nearly identical patterns is well-known in the moiré effect (Fig. 14). Does not something like this operate in the human brain? The generation of the sensation of depth in stereoscopic vision, the sense of beauty that arises in partial but not exact repetition in a painting or landscape, and the emotional effect of counterpoint and harmony in music all come from the superposition in the mind of a newly sensed (or thought) pattern on a structure that was previously present. Perception of

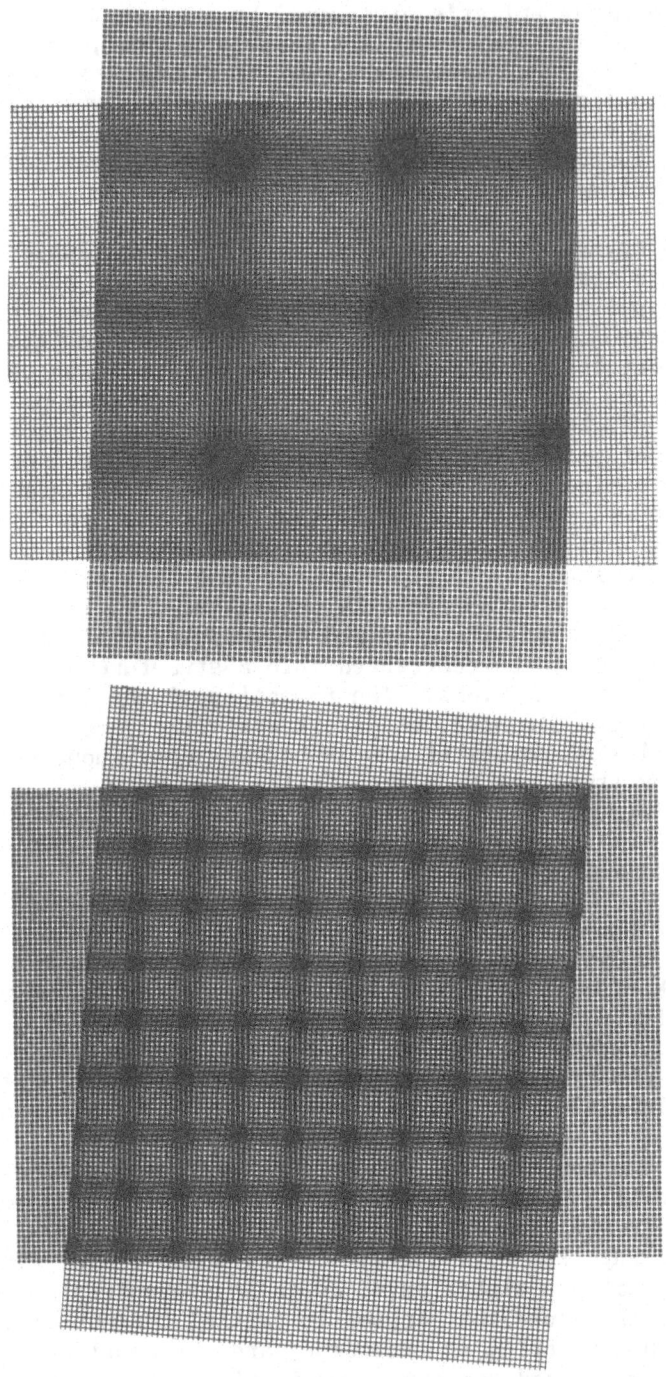

Figure 14

any kind involves only a slight change from what has been
there before. The difficulty of original thought, of nu-
cleating a new chemical phase or establishing a new politi-
cal system are all a result of structural inertia, but the
fact that they are possible at all comes from the suggestions
provided by misfits at the interface between two structural
regions.

<u>On Time</u>

Whether or not a connection between two vertices is actually made is a question of time - or perhaps more properly
the inverse, that time is the sequence of connections. The
changing present is a sort of moiré pattern formed between
the structures of past and future, which are relatively more
stable because they involve more dimensions of interlock.
Time is reversible only in rudimentarily simple structures of
the type favored by the physicist. In these the reversal of
the forces leaves the structure unchanged,-but in complex
systems local geometric interlock makes hysteresis inevitable. Differences between neighbouring groups disperse and
restructure the interactions between the units at their
boundaries.

It was suggested above that polygons and polyhedra do
not "exist" but mark only a defocused view of envelopes terminating smaller groups of linear communication. Even the
lines of communication do not necessarily persist, though
their effects may. A tool confers shape, radiation may produce structural damage, an idea may be transferred via a book
or immunity to disease via a molecule. A message conveyed by
electronics can play just as effective a role in stablizing
or destablizing a structure by the change it produces in the
valence of atoms or minds as does the attachment of wires to
the rods in tensegrity architecture: in fact it was an idea
that preceded the making of the joints in the latter. Stable
structures are resonant ones, ones in which exchange of mes-
sages (ultimately photons in things on our scale of the
world) reinforces internal structures by their emission and
return, not necessarily by the same routes.

An organism is dynamic, it is distinguished from an aggregate in that, superimposed on the necessary 3-D basis of
support, there are a number of local special structures
which contribute to the functioning of the whole via a system of specific communications involving smaller structures
moving both ways but not necessarily maintaining the same dimensionality or returning in comparable times. Communication
may be electronic, sequential and directable through channels
or it may be by broadcast diffusion depending simply on

chance contact as of a scattered seed.

A viewpoint such as the present is intended to be quite general. However, it is useless unless the vertices and the connections to which it is to be applied in any particular case have been identified. Without knowledge of all connections that can influence the structure and behaviour of the units at any level quite absurd conclusions about the whole can be reached. Interest will almost always inhere precisely in those things that are not easily perceived. Complex communication is usually handled better by the methods of the artist than by the scientist's traditional logic. Perhaps, however, a hierarchical view of structure and connections may help a little to bridge the gap.

Acknowledgment

The author is grateful to Professor Arthur Loeb, a fellow Philomorph, for general discussion of the topic of this paper and for his having called attention to some misleading and incorrect statements in an early draft.

References

1. C. S. Smith, "Art, technology and science: Notes on their historical interaction", in Duane Roller, Editor, Perspectives in the History of Science and Technology (Norman, Oklahoma, 1971, pp 129-165. Preprinted in Technology and Culture, 11, (1970), pp 493-549.

2. C. S. Smith, "Structural hierarchy in science, art, and history", in Judith Wechsler, Editor, Aesthetics in Science, (Cambridge, Mass., MIT Press, 1978), pp 9-53.

3. J. L. Locher, Editor, The World of M. C. Escher, (New York, H. N. Abrams, Inc., 1971), Plate 181.

4. Philip Morrison, "On broken symmetries", in Judith Wechsler, Aesthetics in Science (Cambridge, Mass. 1978)pp 55-70.

5. Arthur L. Loeb, Space Structures, their Harmony and Counterpoint. (Reading, Mass., Addison Wesley Publishing Co. 1976), pp 23-27.

6. A. F. Wells, "The geometric basis of crystal structure", Parts 1 and 2, Acta Crystallographica (1954), 7,535 - 54 and subsequent articles in volumes 8, 9, 16, and 18.

7. John G. Burke, The Origins of the Science of Crystals. (Berkeley, Cal., University of California Press, 1966).

3

Algorithms, Structures and Models

Arthur L. Loeb

There are those among us who labor under the delusion
that computer art is a by-product of computer science and
technology. The truth, however, is the exact opposite,
for Charles Babbage's nineteenth century designs for calculating automats were based on the principle of the
Jacquard weaving loom. This loom, having as input a program punched out on cards, is an ancestor of all programmed
machines, computers as well as graphic-design automats.
This example is one of many that bear out Cyril S. Smith's
contention that artistic needs provide the principal moti-
vation for technological innovation and scientific inven-
tion.

Although the Jacquard loom accepts all information
necessary for producing a pre-set pattern on punched cards,
it has no means of altering the program. In this respect
twentieth century computers have a greater degree of sophis-
tication, for they possess an internal memory whose content
may be changed by the computer itself. The Turing [1] machine
is an extreme example of a minimally implemented device
which may organize itself into a very sophisticated perfor-
mer under control of a minimal read-in program.

Biological processes, too, may occur as a consequence
of a minimal pre-set program, for instance, a genetic code.
It is not surprising, therefore, that electronic as well as
biological-systems analysts have come to study those ger-
minal codes or functions which, frequently by means of
feed-back control or internal self-regulation generate
very complex structures or patterns. Such generating
functions are called algorisms or algorithms. We shall see

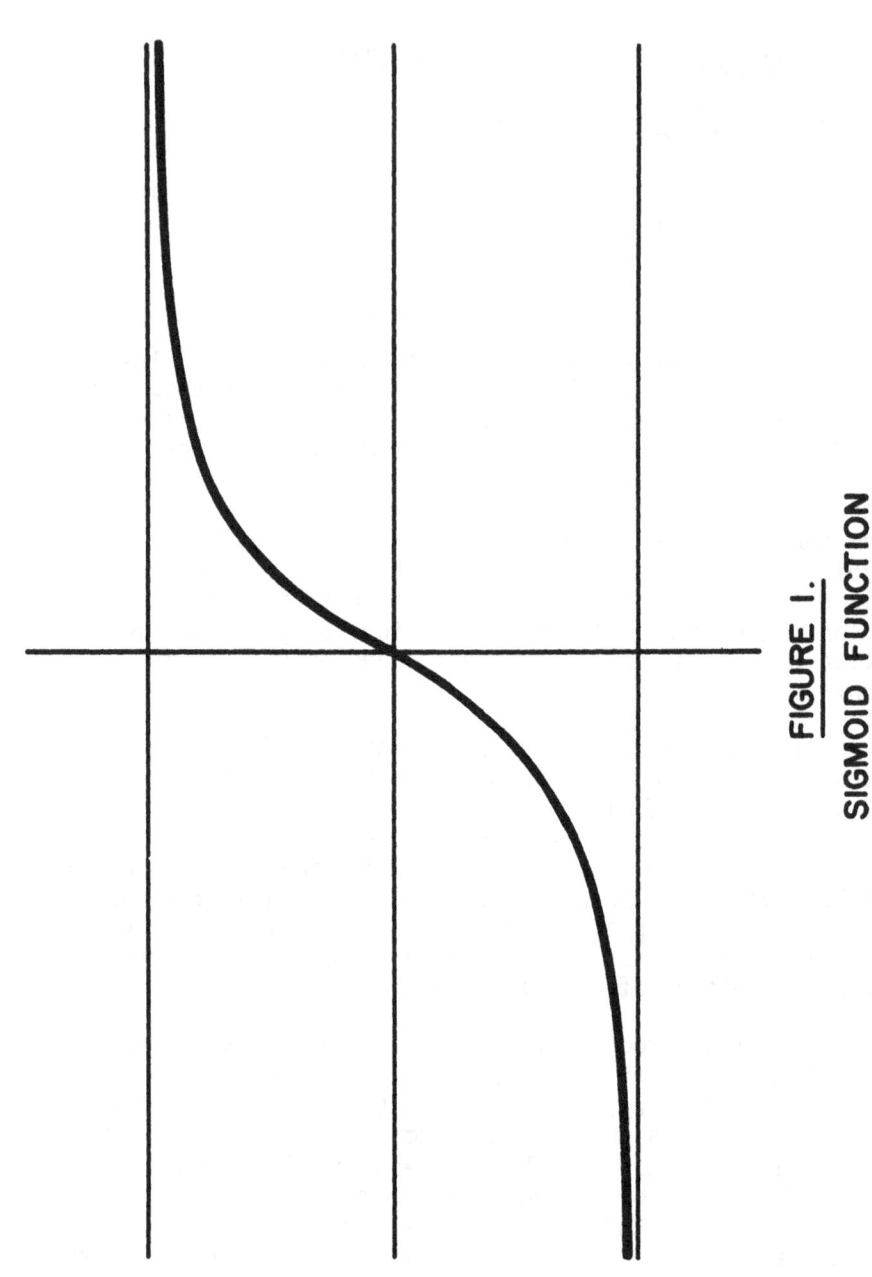

FIGURE I.
SIGMOID FUNCTION

that quite frequently our perception of a structure or pattern is altered appreciably when we approach the pattern algorithmically, that is to say, from the point of view of the algorithm which generated the pattern.

As a first example, let us consider so-called growth functions. These are functions whose rate of change with time is not explicitly dependent on time, but only on the value of the function itself at a given moment. Such a function is, after a given starting condition, totally self-generating, and independent of external events. Its growth is generally described by a differential equation of the form $\frac{dy}{dt}= f(y)$. The simplest expression for the growth

rate is, of course, $f(y)$ = constant; the function y then changes linearly with time. Next would come a function $f(y)$ which is linear in y; it corresponds to an exponential change of y with t, applicable as much to the growth of capital subject to continuously compounded interest, as to the growth of the Nautilus shell.

Familiar though we may be with linear and exponential growth, we tend to be baffled by the saturation phenomenon which appears to limit many growth phenomena in the long run. It was fashionable in the early nineteen sixties to comment on the apparent exponential growth of the number of natural scientists with the remark that by the year 2000 A.D. there would be more scientists than people. In spite of the obvious absurdity of this prediction, the scientific community was appalled when the exponential growth did not continue.

Derek de Solla Price [2] has noted that a great many growth processes that were monitored over long periods of time, such as the number of universities in Europe, appear to run through three phases: an initial phase of very slow growth, followed by a phase of very fast growth, and con- cluded by a phase in which growth gradually comes to a standstill. Such growth processes are characterized by a so-called sigmoid curve (Figure 1). It is interesting to speculate whether these three phases are controlled by changes in external conditions, or whether a single algo- rithm might in fact account for all three phases. In the latter case the saturation phenomenon would be inherent in the growth mechanism, and might be predictable even in the early stages of growth.

Let us return to our growth functions, and consider a quadratic form of $f(y)$, in particular one having two real roots. By proper scaling of variables we can, without loss of generality, set f(y) = k(1-y^2), so that

$$y = (e^{kt} - e^{-kt}) / (e^{kt} + e^{-kt}) \equiv \text{tank } kt$$

This function approaches -1 when $t \to -\infty$, and +1 when $t \to +\infty$; the asymptotic values y = $\pm$1 represent no growth $\left(\dfrac{dy}{dt} = 0 \right)$.

Maximum growth occurs when y = 0. In fact, this function is exactly the sigmoid curve shown in Figure 1. There are, in point of fact, no three separate growth phases, no discontinuous alterations, and no need for external variables to account for saturation. Like linear and exponential growth, sigmoid growth is algorithmical. We observe therefore, that the perception of sigmoid growth as being constituted of three separate phases is actually altered by considering the algorithm $\dfrac{dy}{dt} = k (1-y^2)$.

A pattern is an ordered array. Structure is a set of relationships between the components of a pattern. A structural model is an analog of a pattern, which incorporates a subset of the structural relationships of the pattern. The model is frequently scaled up or down; it is also necessary in some instances to distort distances in order to preserve topological relationships or laws of motion. Usually a structural model is a valid analog of the original pattern only over a restricted order of magnitude. Consider, for instance, the structural model in Figure 2, which is a representation of the mineral spinel. Every distance in the original pattern has been scaled upward by a factor of approximately 10^9. The model has applicability only to structural relationships ranging over distances (in the original) of about 10^{-7}cm, and makes no claim to represent relationships within ions. The function of the rods in the model is merely to hold the spheres in place; they do not represent significant relationships in the original pattern. The model is predicated upon the spatial relationships that affect the scattering of X-rays, but it is not very effective when, for instance, magnetic interactions in spinel-like materials are to be investigated. Consider, by contrast, Figure 3, which actually models the same spinel structure in the same magnification as did Figure 2. This model was generated by the following structural algorithms:[3]

1. Each oxygen ion is equidistant from as many other oxygen ions as possible. (The oxygen atoms are, in this model, represented by the vertices of the tetrahedra and

octahedra).

2. Each magnesium ion is equidistant from exactly four oxygen ions, and equidistant from as many other magnesium ions as possible. (Magnesium ions are represented by spheres).

3. Each aluminum ion is equidistant from exactly six oxygen ions, and equidistant from as many other aluminum ions as possible.

4. As many aluminum-magnesium ion distances as possible are equal to each other.

5. The chemical composition of spinel is $Mg\ Al_2\ O_4$.

These algorithms derive from a more general algorithm, the Vector Equilibrium Principle, discussed elsewhere.[4,5] The structural model of Figure 3, being based on these algorithms, naturally displays these geometrical relationships most prominently, and as a result, is very suitable for tracking magnetic interactions within spinel-like ceramic materials.

Models based on structural algorithms serve to show probable and improbable configurations. Thus they are help-ful in the prediction and elimination of certain configura-tions, and in turn in the evaluation of experimental results, which are much more easily interpreted when one has a notion about where ions are or are not likely to be situated. Such models are also used to study imperfections, for instance when the chemical composition of a material is not exactly stoichiometric. We have thus demonstrated how two different "Gestalten", namely Figures 2 and 3, of the same configura-tion can serve entirely different purposes.

We tend nowadays to model ions as spheres. In some models, e.g. that of Figure 2, these are held in place by rods. In other models the spheres are made tangent to each other, their various radii corresponding to the so-called size of the ions. It has not always been so. In the early nineteenth century the polyhedral shapes of large crystals led scientists to imagine these crystals as built up from polyhedral modules of near-atomic scale. The question which model is "correct" means little when we recall that we are modeling configurations whose scale is but a fraction of the wavelength of visible light, and hence beyond the range of our visual experience. What we do know from physical exper-iments is that atoms and ions in crystals have their mass concentrated within a radius of the order of 10^{-12}cm, but that these masses are separated by distances about 10,000 times as large. A model showing the distribution of mass

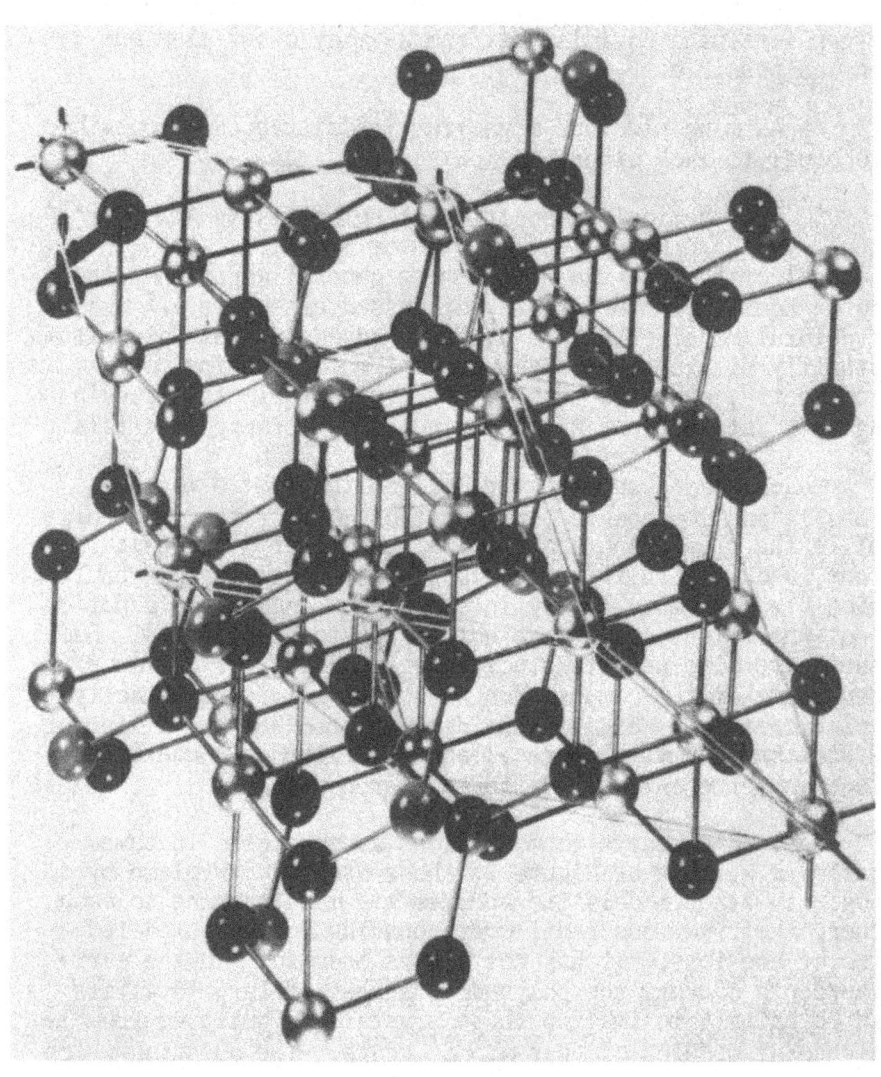

Figure 2. Ball-and-rod model for spinel

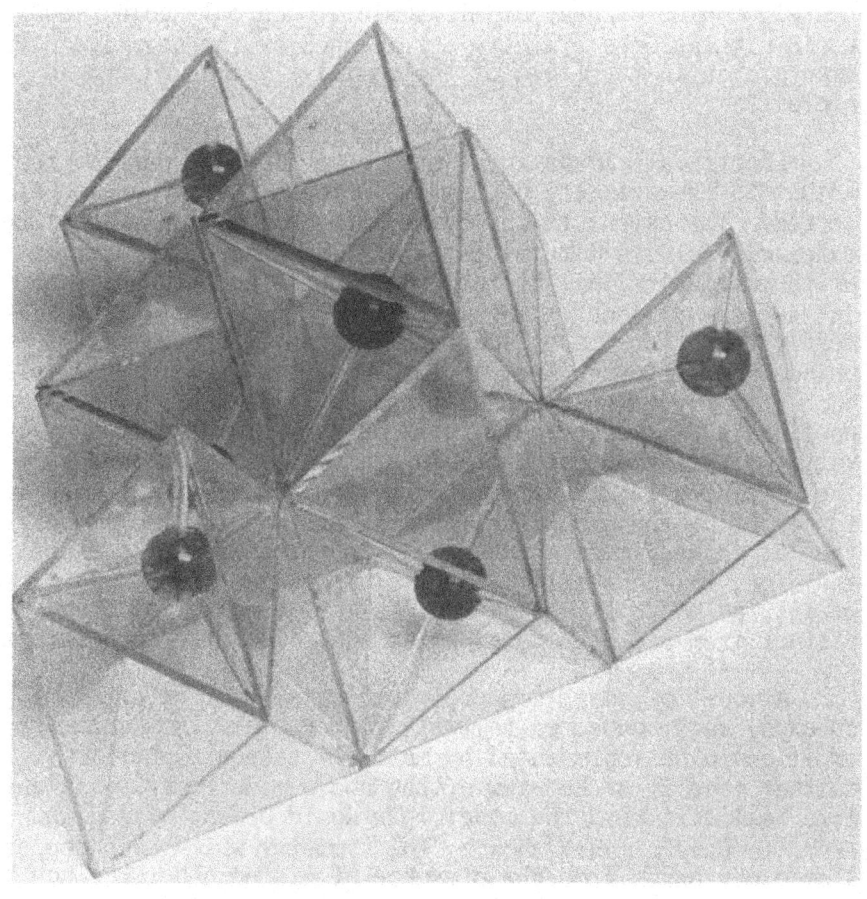

Figure 3: Moduledra model for spinel

in a crystal would therefore need to consist of very tiny
spheres, separated by enormous distances. On the other hand
a model for the distribution of electrical charge would
have positive charge concentrated where the mass is, but
would have negative charge distributed over the entire space
in between, local charge density being determined by the
symmetry of the positive charge distribution. Such an elec-
trical-charge model would therefore resemble that of polyhedral modules of near-atomic scale. Such polyhedral modules
would simulate the symmetry experienced by the negative electron clouds in the field of the positively charged atomic
nuclei.

Electrical clouds in an atom have no shape per se: they
adjust to the symmetry of the field to which they are sub-
jected. The extent to which they are able to conform to the
external field is determined by the internal configuration
of the atom (or ion), primarily by the number of electrons
in the atom or ion. A sphere as model for an atom signifies
merely that there is no single direction in space that is
unique and different from any other direction. In principle
each atom may be surrounded by as many equivalent ions as
possible. The upper limit to the number of mutually equivalent points that may be equidistant from a given one of
their members is not determined by the nature of interatomic
forces or by the properties of a sphere, but by the properties of three-dimensional space. If the distance be exactly
the same, this upper limit is twelve; if we permit a variance
of 15%, as many as fourteen points will be able to be equidistant from one particular point equivalent to each of them.

A model of, say, copper, could consist of an array of
spheres, each connected to twelve other equivalent spheres,
or it could be represented by mutually tangent spheres.
Neither model, in the view of the above-made remarks, is any
more realistic than the other. It would be satisfying, however, to have a model showing the symmetry which each copper
atom experiences from the presence of its neighbors. Such a
model would represent each copper atom by a polyhedral cell;
when these cells are packed together to fill space, the resulting structure is the dual of that constituted of the
spheres connected to each other. The dual of a three-dimensional configuration consisting of vertices, edges, faces and
cells is obtained by replacing each vertex by a cell, each
edge by a face, each face by an edge, and each cell by a vertex. [In general, the dual of an N-dimensional configuration
is found by replacing every k-dimensional element by an (N-k)-
dimensional one. Polyhedral surfaces are 2-dimensional,
hence vertex is replaced by face, edge by edge, and face by
vertex: the dual of a cube is an octahedron, etc.]

The polyhedral model of copper would consist of packed rhombic dodecahedra. Each atom would share a face of its twelve neighbors. For a so-called body-centered metal, in which each atom is surrounded by fourteen equivalent approximately (tolerance 15%) equidistant neighbors, the cell is a truncated octahedron having thirty-six edges of equal length. Eight nearest neighbors share hexagonal faces with each atom, while six slightly more remote neighbors share square faces with each atom. (Figures 4 and 5).

We see thus that for every spatial distribution of discrete entities there are two mutually dual representations. One represents each entity by a point of space; edges connecting these points show the interconnections between them. Its dual representation assigns to each entity the entire space proper to it: the interconnecting edges then become interfaces. The cells in such a model are called Dirichlet Domains: properly defined a Dirichlet Domain of one of a discrete array of points is the innermost cell whose faces perpendicularly bisect the connections from that one to all the other points. (It is in real space what the innermost Brillouin zone is in reciprocal space). When all of the points of such an array are mutually equivalent, all Dirichlet Domains are mutually congruent. Thus to every crystallographic lattice there corresponds a space-filling polyhedron, which is its Dirichlet Domain. The Dirichlet Domain of the "cubically close-packed" (also known as "face-centered cubic") lattice occupied by the above-mentioned copper atoms is the rhombic dodecahedron, that of the body-centered lattice is the truncated octahedron.

Faces of Dirichlet Domains are equidistant from two of the array of discrete points. Each edge of a Dirichlet Domain is equidistant from at least three points, each vertex from at least four points. The dodecahedral Dirichlet Domain proper to the face-centered cubic lattice has two different types of vertices. At one such type four acute angles meet: such vertices are equidistant from six lattice points. At the other type of vertex three obtuse angles meet: such vertices are equidistant from exactly four lattice points. If we place oxygen ions at the centers of the dodecahedral Dirichlet Domains, these ions are equidistant from twelve equivalent ions, the maximum number allowed. They therefore satisfy the first of the algorithms required above for the spinel structure. Magnesium ions symmetrically distributed over the obtuse vertices, and aluminum ions symmetrically distributed over the acute vertices satisfy the remaining algorithms. Thus a hierarchical model is arrived at, corresponding to models first designed by E.W. Gorter[6] .

We see thus that Dirichlet Domains provide models dual

Figure 4. Rhombic dodecahedron

Figure 5. Space filling with truncated octahedra

to the usual ball-and-rod models, and also that they provide
a hierarchical approach to structure: the vertices of the
Domains provide likely locations for a second type of ions.
These vertices are, in fact, generalizations of what, in cubically close-packed structures are called "interstices".
The author has shown elsewhere[5] that a double-layered hierarchy may be derived from the body-centered lattice, account-
ing for such structures as β- tungsten and β- uranium hydride.

In two-dimensional planning and design, too, the Dirichlet Domain may have its uses. Such randomly placed features of the urban scene as churches, schools and subway
stops have their parishes, districts and domains. Construction of Dirichlet Domains on a city map instantly reveals to
any inhabitant the church, school or subway stop nearest to
his or her home or place of work. And in the design of
tessellated surfaces for buildings or street surfaces, algorithms using Dirichlet Domains provide an interesting repertoire using only very few different shapes.

A plane may be filled by mutually congruent triangles of
any shape. The algorithm for filling the plane is that the
center point of each edge of each triangle be a center of
two-fold rotational symmetry. In order for a triangle to be
not only a plane-filler, but also a Dirichlet Domain, there
must be a point inside the triangle that is equidistant from
each of its three vertices. This is true only if none of the
angles of the triangles exceeds 90°. In this case the array
of points proper to the triangular Dirichlet Domains constitutes the centers of circles drawn through the vertices of
each triangle.

A plane may likewise be filled by mutually congruent
quadrilaterals of any shape; again the midway points of each
of the edges must be centers of two-fold rotational symmetry.
In order for a quadrilateral to be a Dirichlet Domain, its
vertices must lie on a common circle. In two dimensions the
vertices of Dirichlet Domains and the points at the centers
of the circles circumscribed around the Domains have an interesting interrelationship. If we call the set of original
points A, and the vertices of their Dirichlet Domains B, then
the Dirichlet Domains proper to the set of points B have as
vertices just the set of original points A. Moreover, the
network formed by the Dirichlet Domains proper to the points
A is dual to that formed by the Dirichlet Domains proper to
the points B. For instance, in Figure 6 the points A are
centers of triangles whose vertices are the points B, which
in turn are the centers of hexagonal Domains whose vertices
constitute the set of points A[7] .

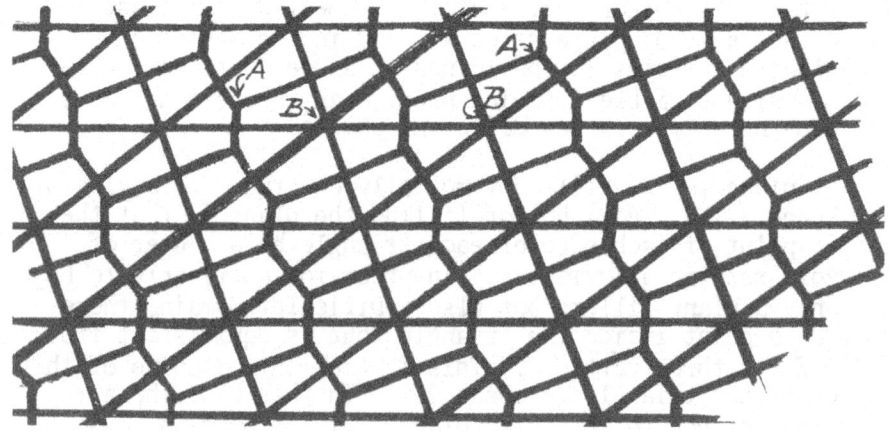

Figure 6: Triangular and hexagonal Dirichlet Domains.
A̲ marks vertices of Domains proper to points B̲,
while B̲ marks vertices of Domains proper to points
A̲.

Figure 7 shows a set of points A whose Dirichlet Domains are quadrilaterals having as vertices the points B. In turn, the points A are the vertices of the Dirichlet Domains proper to the points B. If two adjacent angles of a given Domain are called α and β, then the angles diagonally opposite them are respectively (180°-α) and (180°-β). The dual net must have the same four angles, but is generally not congruent with the original net because the edge lengths are not correspondingly equal. If, however, either α or β equals 90°, then we encounter a special kite-shaped Dirichlet Domain which is congruent with its Dual Domain. In this particular instance the two dual nets subdivide each others' faces into four sub-quadrilaterals, of which two are geometrically similar to the original quadrilaterals, the other two rectangular (Figure 8). This phenomenon permits a variety of tessellations using three basic shapes, two rectangular, the third quadrilateral, having only two right angles diagonally opposite each other.

In summary, then, we have observed that our perception of an apparently complex configuration is altered when, instead of attempting a complete description of the object, we generate the configuration from a small number of relatively simple modules together with an algorithm for assembling them. This new perception resolves the complexity by a hierarchical process, which resolves the structure into, as it were, separate but superimposed layers (e.g. Figure 8 as generated from Figure 7).

Generally, we do not know the modules and algorithms which would generate a given complex configuration. The role and process of science would seem to consist of a search for appropriate modules and algorithms which generate models whose behavior resembles adequately that of the complex configuration studied. The analogy between model and observed configuration is limited and quite subjective, depending on the observer, and the purpose, context and background of the experiment.

In design, the algorithmatic approach generates with simple means a rich repertoire of patterns transcending the repertoire of the "naked eye". In addition, the conceptual component of such a generated pattern has an esthetic appeal of its own, and constitutes an important link between art and science.

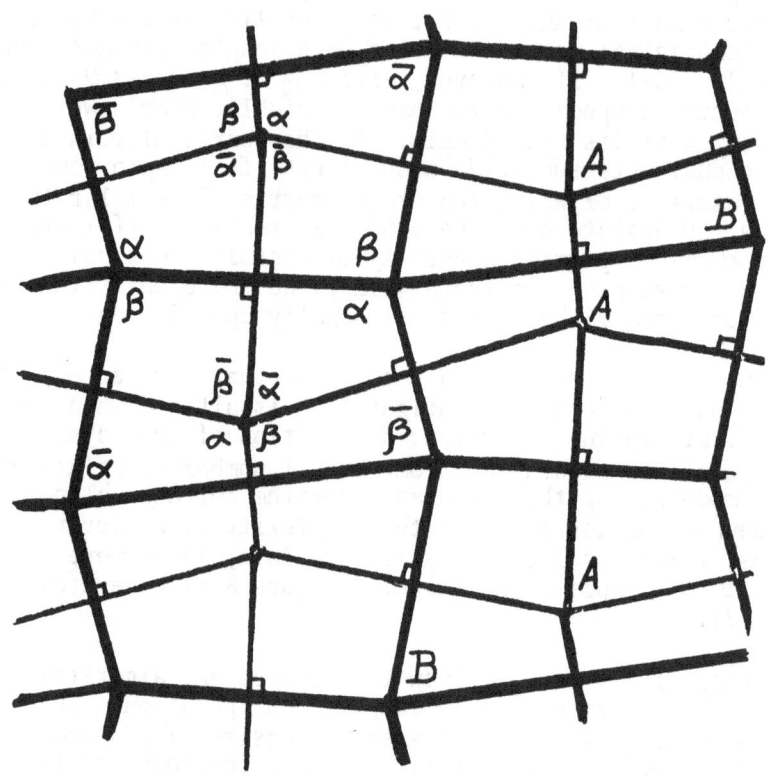

Figure 7. Quadrilateral Dirichlet Domains. A marks vertices
of Domains proper to points B, while B marks vertices of Domains proper to points A.
$\bar{\alpha} \cong 180° - \alpha$; $\bar{\beta} \cong 180° - \beta$
The two dual sets are mutually congruent only when
α or β equals 90° .

Figure 8. Design using superposition of mutually dual and congruent Dirichlet-Domain nets. The four basic modular shapes are emphasized.

References

1. Called after A. M. Turing: cf. "Systems of Logic based on Ordinals", C. F. Hodgson and Son, Ltd., London (1939).

2. Derek J. de Solla Price: "Measuring the size of science", Proc. Israel Academy of Sciences and Humanities, Vol. 4, Number 6, Jerusalem (1969).

3. A. L. Loeb and G. W. Pearsall: "Moduledra Crystal Models" Amer. J. Physics, 31, 190-196 (1963).

4. A. L. Loeb: Contributions to R. Buckminster Fuller's "Synergetics", pp. 860-875, MacMillan, New York (1975).

5. A. L. Loeb: "A systematic survey of cubic crystal structures", J. Solid State Chem., $\underline{1}$, pp. 237- 267 (1970).

6. E. W. Gorter: "Some structural relationships of ternary transition metal oxides", XVIIth International Congress of Pure and Applied Chem., pp. 303-328, Butterworth's, London (1960).

7. A. L. Loeb: "Space Structures, their Harmony and Counterpoint", p. 120, Addison-Wesley, Reading, Mass. (1976).

4

Complex Relations in
Urban and Regional Planning

An Application of Hypergraphics

C. Ernesto S. Lindgren

Introduction

Internal structure of cities, urban economics, regional development, land use theory, urban invasion and succession, transport planning, housing, cities systems, urban renewal and several other topics pertaining to urban and regional planning require, each, the consideration of a great number of variables which the analyst combines or tries to combine, through algorithms, models and other mechanisms, in order to achieve some comprehension of their inter-relations, interplay, mutual influences, etc. In all, the experience is quite unsatisfactory as a result of two basic deficiencies: (a) lack of comprehensive urban and/or regional planning theories; (b) lack of know-how in the building of mathematical frameworks isomorphic to that structured by the variables. I am not certain which of these two deficiencies is more critical but we know that they are related, meaning that an advance towards resolving one problem will certainly facilitate solving the other. I came to the conclusion, however, that thinking in terms of hyper-graphics appears to help the understanding of the significance of a certain event, localized in a hyperspace through coordinates that represent the state of that event, at a certain time and place. In this paper, I propose to share with the reader some of the experience that I have had in dealing with complex relations in urban and regional planning through application of hypergraphics.

The Basic Problem

Assuming a two-variable problem, suppose that we have a set of events, each event characterized by a pair of variables, setting aside time and geographical location of the event. We can also think of one single event characterized by pairs of variables, changing through time.

Each event (or each set of two variables) can be represented graphically, in a Cartesian system. Applying a numerical method, an equation can be derived which then could be interpreted as the model for the history of the set of events or for the history of the set of variables associated with the event. At this point, nothing has been elaborated about a theory explaining the variations. We are concerned only with the representation of that variation. All is well, so far, because: (a) we can count on that Cartesian system to represent the variables, that is, they are made isomorphic to each of the two axes; (b) we have methods to derive some kind of curve that is then isomorphic to the structure defined by the series of events; evidently, as we choose the curve (the degree of the polynomial adjusted to the scatter of points in the plane) we assume that it is the best representation of the isomorphism. Supposing however that we have the means to operate with the Cartesian system or any other kind of two-dimensional representation but that we do not know how to describe isomorphisms between geometric forms in that two-space and the history of the event. The basic problem would then be the need to propose methods of adjusting polynomials to a scatter of points in the plane. An absurd supposition, one would think but this is the situation that planners must face when dealing with complex relations involving several variables. The isomorphism between variables and axes of an n-dimensional space has been resolved but the isomorphism between the scatter of points in n-space and a geometric form of proper dimensionality has yet to be devised.

A first step towards the solution of this dilemma is the generation of relative indexes in place of the absolute or even the percent value of a variable. This has been achieved with the use of relative indexes of concentration, segregation, dissimilarity, etc.

Having attained the isomorphism between variables and axes of an n-dimensional space and obtained the location, in n-space, of each point, two questions are asked by the planner: (1) what is the result of the combination of those n - variables, meaning that one seeks to measure the combined effect of the variables so that the event to which they correspond is unique at a certain time and place; (2) how to describe the history of a succession of events, meaning that one seeks to adjust some geometric form in n-space to a scatter of points.

In my research I have used three basic forms of dealing with the questions, as briefly described below. The underlying reasoning is that, given a set of variables, we are

concerned primarily with the combined effect of their
presence, secondarily with the interrelations, inter-
dependencies, etc. This does not mean, of course, that the
relations among variables is not significant. The point,
however, is that it becomes more fundamental to identify the
simultaneous influence of dependent and independent
variables rather than their degree of correlation or
association.

First Method

A straightforward approach is to consider the
variables or the relative indexes that represent them as
vectors acting at a certain point. The combined effect of
these vectors is the resultant, that is, the diagonal of a
hyper-prism constructed in n-space, where n is the number
of variables.

This form of representing the resulting effect of
several variables has been used, with some degree of success,
in problems where classification is used in order to
differentiate geographical points and make value judgments
as to the outcome of the combination of two or more
variables and not of their degree of association. For
example, we might be interested in identifying those
geographical points in a city, for instance residential
areas, more vulnerable to urban invasion and succession. At
each point, variables such as "pressure from real estate
agents" and "family demand as an indication of desire of
achieving higher social status" might be identified and
properly measured. If our objective is to verify which
families are more likely to move than others, to some other
geographical point, it is more significant to indicate how
those two variables, acting simultaneously, generate an
influence strong enough to cause the move or not. In this
case, measurement of the degree of association of the two
variables, that is, determining to what extent "family
demand as an indication of desire of achieving higher social
status" is explained by "pressure from real estate agents"
influencing family members is not as pertinent as it might
be considered. At a certain moment it must be considered
that these two variables do act together, summing their
effects, regardless of the fact that they are related or not.

As a way to produce indicators to be used in a
classification of counties by land use type, the method
yielded results that were matched against the scores in each
of three factors obtained through factor analysis (orthogonal
rotation) involving over forty variables. Over 150 counties
were considered and the correlation coefficient (product-

moment of Pearson) was consistently above the 0.90 mark. In this case, instead of working with the z-scores of the variable, they were transformed in relative concentration indexes (Lindgren, 1975).

These two examples of application of this method, which dispenses assumptions about the kind of statistical distribution of each variable and permits the standardization of measurements at different scales (nominal, ordinal or ratio) through the estimation of indexes of concentration of the variables, appears to be a useful contribution to the understanding of such complex problems.

Second Method

As in the above method, all variables of a given problem are standardized as relative indexes of concentration and we seek to represent the final geometric form that n variables yield in an n-dimensional space. Since we are consistently operating with areal distributions, the geographical dimensions and any one variable define a continuous surface in 3-space. Next, a second variable is represented orthogonally to this 3-space, generating a continuous hypersurface in 4-space. And so on.

The method of obtaining these spatial relations is the descriptive geometry method which has been developed and extended to any dimension (Lindgren and Slaby, 1968). At the moment, all constructions are done manually, a restriction that does not permit the application of the technique to more than five or six variables (6-D descriptive geometry method). The interpretation of the relations among variables, two by two or three by three, etc., is immediately obtained by examining the projections of the geometric form in n-space on the (n - 1)-spaces of the system of reference for the descriptive geometry method.

I can report only one application of this technique and that was in the case of generating an algorithm for the graphical representation of a matrix of social distances (Lindgren and Steinitz, 1969).

Third Method

A model of spatial distribution and flow (Lindgren, 1969) has been proposed for areal distributions that are embedded in 3-space. Usually, two of the dimensions are used to locate the geographical position where a certain phenomenon is measured. It was then suggested that, since more than one variable is measured at each geographical

location and these variables, associated or not, would be
related to the same problem under examination, the proposed
model should be adapted to take into consideration this
multi-dimensional aspect. This adaptation has not, as yet,
been accomplished by virtue of technical difficulties, as
follows, and should perhaps become of interest to some
scholar.

In the proposed model (3-D version) flows among
geographical locations are suggested to occur along the
geodesic lines connecting any two locations at their
corresponding positions on the 2-D surface that represents
the areal distribution. The distribution of a certain factor
(population, income, industrial output, etc.) among these
locations are suggested to be influenced by the length and by
the slope of the geodesic lines. Since the mathematics of
the 2-D surfaces that are generated is usually intractable,
the location of the geodesic line is obtained by a graphical
method (Lindgren, 1970).

As we proposed to adapt this model to more than one
variable with distributions over a geographical area, by
suggesting that flows and distributions are to take place and
be influenced by the geodesic line, it will be necessary to
develop a way to rework the problem by considering 3-D spaces
in place of the 2-D surface. This means that given a 3-D
space or, more exactly, a hypersurface, necessarily with
curvature and other corresponding properties to those of a
2-D surface, it will be necessary to state what is the nature
of these properties and what are the parameters to be
measured at each point. One would need to define that
geometric form which, in that hypersurface, corresponds to
the geodesic line on a 2-D surface.

It is apparent to me that these demands require
considerable work in the area of curved hypersurfaces
generated in 4-D and higher-dimensional Euclidean spaces. In
particular, just as curvature at a given point of a 2-D
surface is measured, the definition of curvature at a point
and at a line in 3-D curved surfaces must be defined. In the
face of these shortcomings it is also apparent to me that the
suggested adaptation of the above model is unlikely to occur.

Conclusion

Briefly, these have been my experiences in the
application of hypergraphics in complex problems in urban and
regional planning, a field where I have been doing research
for only five years. It is highly probable that other
successful applications, similar to that case of the high

correlation with the factor analysis scores, will be achieved. Because of the classificatory nature of the problem there is no risk in affirming that a vast number of problems could be managed in the same form.

References

Lindgren, C. Ernesto S. and Steve M. Slaby, Descriptive Geometry of Four Dimensions, New York: McGraw-Hill Book Co., 1968.

Lindgren, C. Ernesto S., "Descriptive Geometry of Four Dimensions" (collection of papers from the Engineering Graphics Seminar), Princeton: Princeton University, 1963-1964.

Lindgren, C. Ernesto S., collection of papers prepared for the series on Theoretical Geography, Cambridge: GSD/ Harvard University, 1967-1972.

Lindgren, C. Ernesto S., "The Design of a Deterministic Model of Spatial Distribution and Flow" (unpublished master's thesis), Medford: Tufts University, 1969.

Lindgren, C. Ernesto S., Análise de Dados em Planejamento Urbano e Regional, Rio de Janeiro: COPPE/ Universidade Federal do Rio de Janeiro, 1975.

Lindgren, C. Ernesto S., Developments in Four-Dimensional Geometry, Rio de Janeiro: COPPE/Universidade Federal do Rio de Janeiro, 1977.

5

Seeing Order

Systems and Symbols

Anne Griswold Tyng

We see as a result of how we are made. A similar geometry orders natural form and human perception. An order
based on a sequence of geometric principles underlies the evolution of natural forms. The same ordering sequence appears
in architectural history as the underlying geometry of changing 'styles'. What has been called 'styles' of architecture
occur as the result of shifting phases of form empathy in human consciousness. As an extension of the evolution of natural forms, the evolution of human consciousness follows the
same geometric ordering system. Out of these fundamental
patterns of perceiving, the human spirit has transformed the
geometry of natural systems to symbols in architecture, in
science and in art.

In 450 B.C., in his search for an 'atomic' order of spatial concepts, Empedocles proposed as the building blocks of
everything the four 'elements' fire, air, earth, and water.
On mathematical grounds Plato, in his *Timaeus*, determined the
'exact' forms of the smallest parts of these elements as the
five shapes we now call the Platonic Solids; fire the tetrahedron, earth the cube, air the octahedron, water the icosahedron and as the symbol of the cosmos, the dodecahedron.
This intuitive concept is given a measure of validity today
when we know that the relationships of form expressed in
these five Platonic Solids are involved in the way in which
'fundamental' particles -- protons and neutrons -- are built
up into atoms of about a hundred different elements (according to Pauling's Close-Packed-Spheron Theory and Fuller's
proposals of atomic close-packing) (1) and are involved in

---

[1]Buckminster Fuller lectured on and illustrated relation-
ships of close-packed spheres as proposals of atomic configu-
rations in 1949, printed as Item O prepared by North Carolina
State School of Design students in 1955. Linus Pauling's
Close-Packed-Spheron Theory of the Nucleus appeared in
Science, October 1965.

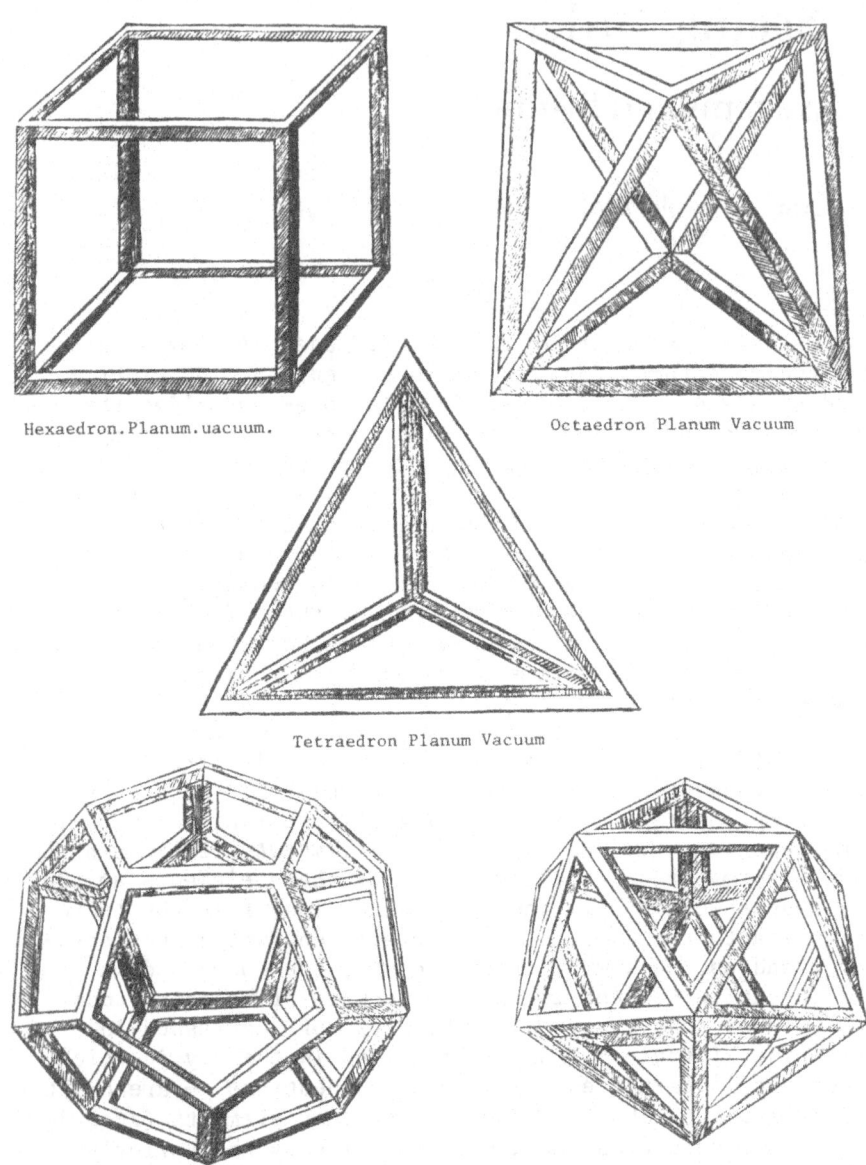

Hexaedron.Planum.uacuum.

Octaedron Planum Vacuum

Tetraedron Planum Vacuum

Dodecaedron Planum Vacuum

Icoſaedron Planum Vacuum

The Five Platonic Solids drawn by Leonardo da Vinci
from Pacioli's <u>De Divina Proportione</u>

the way in which different arrangements of these atoms form
the building blocks of a million or so different forms of
matter, both natural and synthetic.

These five Platonic Solids -- the only regular forms
possible in three dimensional space, each with all of its
faces the same and with the angles at which the faces meet
each other the same -- are involved, not only in the spatial
organization of forms at the level of nuclei of atoms and
molecules, but also in cells, organs, plants, animals, the
human embryo, the psychic structure of man, the works of man
and in the astronomical forms of the universe which pre-ex-
isted man. Previously invisible ordering of the primordial
atoms within us, revealed by the electron microscope, gives
proof of *internal geometry in natural forms,* while recent
psychological insights suggest instinctual images of the unconscious mind as the *profound biological roots of man-made
forms.* Henri Focillon wrote,

> "To assume consciousness at once is to assume form.
> Even at levels far below the zone of definition and
> clarity, forms, measures, and relationships exist.
> *The chief characteristic of the mind is to be
> ceaselessly describing itself.* Forms mingle with
> the life from whence they come; *they translate into
> space certain movements of the mind."* 2 (my italics).

Clues to the geometry of mind-matter exist in fundamental
principles of three dimensional form. In studying the five
Platonic Solids and the relationships between them, I have
found a geometric *progression from simplicity to complexity
of symmetric forms linked by asymmetric process.*

In building the five Platonic Solids one upon the other
in the order of their complexity and in constructing their
spatial extensions, I have found a sequence of four characteristic orders. The sequential order of this underlying geometry of systems and symbols is represented in its simplest
form in the *square,* the *circle,* the *helix* and the *spiral.* Of
these, the most fundamental order is *bilateral* or fourfold
symmetry of the square. In three dimensions it is also seen
in the cube or 90 degree angled orthogonal system, which
finds an ordering source in the human body with a right-left
axis, a front-back axis, and a vertical top-bottom axis. In
architecture, this fundamental ordering system occurs in the
gridiron plans of city streets, in the recurring rectilinear
forms of Greek Temple, of Roman Basilica and the New England
'saltbox'. It occurs in the compass orientation of north,
south, east and west. It appears in our perception of four

2Focillon, H., *The Life of Forms in Art,* published in
Paris 1934, second Edition enlarged. George Wittenborn,Inc.,
1958, pp. 14, 44, and 49.

Major John Bradford House, Kingston, Mass.

Maison Carree at Nimes

forces in the universe, the nuclear force, electrical force,
weak interaction and gravity. This conceptual fourfold
order appears in the four temperaments of medieval physi-
ology, choleric, melancholic, phlegmatic and sanguine, as
well as in the more recent typology of the psychologist Carl
Jung with its four basic orientations of thinking, feeling,
intuition and sensation. The Christian cross is probably
the best known symbol of this order.

The four stages of symmetric form in this geometric
progression, I have called *bilateral, rotational, helical*
and *spiral,* with each stage seen as *the motion of simpler
forms defining the outline of more complex shapes.* The
polarity of a tetrahedron can be expressed in the polariza-
tion of two of its four edges (as Fuller has suggested).
One tetrahedron in two positions, which have a point to face
polarity, can establish the corners of a cube. Two other
positions of a tetrahedron, also in *polarity,* define the
corners of an octahedron. These three simpler Platonic
Solids -- the tetrahedron, cube and octahedron -- represent
the *bilateral* forms of the geometric progression. The cube

POLARITY OF THE SIMPLER SOLIDS

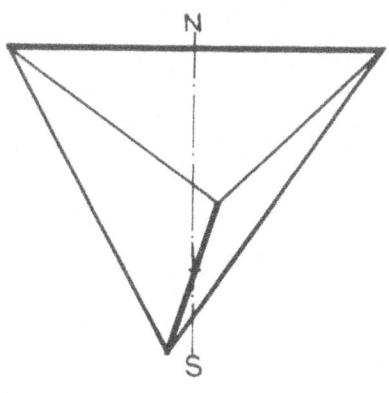

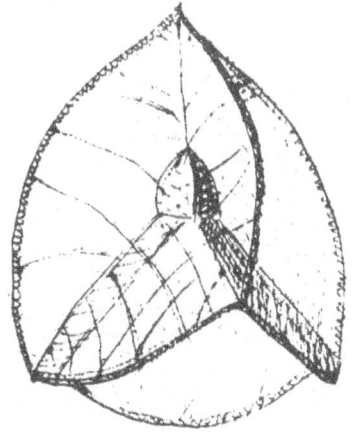

tetrahedron with polarity of
two edges

Righthand drawing is of Nassellarian skeleton
Callimitra agnesae, after Haeckel.

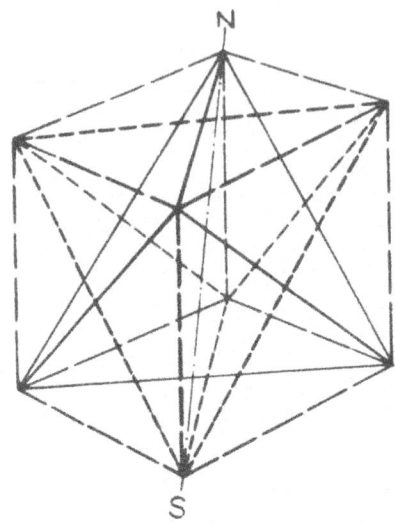

<u>Lithocubus</u> <u>geometricus</u>,
after Haeckel.

cube formed by two positions of
tetrahedron in point to face polarity

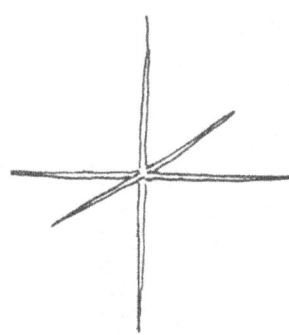

spicule of hexactinallid
sponge, after D'Arcy
Thompson.

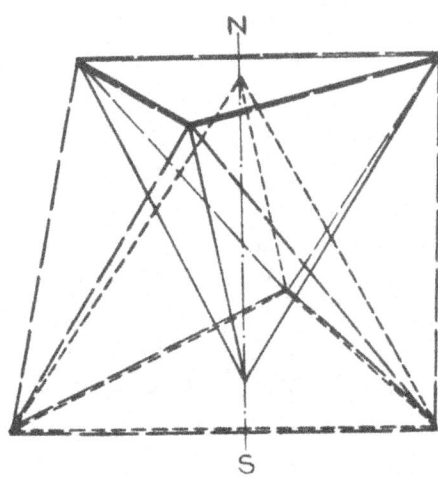

octahedron formed by two positions of
tetrahedron in point to face polarity

in five positions, in *rotation*, defines the twenty corners
of the dodecahedron, and five positions of the octahedron,
again in *rotation*, establish the twelve corners of the icosahedron. The tetrahedron in four positions, with *rotational* ordering, also defines the twelve corners of the icosahedron. These more complex of the Platonic Solids, the dodecahedron and icosahedron, represent the stage of *rotational* forms in the geometric progression, and, in the way they
are formed, express Divine Proportion ratios (1:1.618) in
their relation to the simpler solids, the dodecahedron to
the cube and the icosahedron to the octahedron.

ROTATION IN THE HIGHER SOLIDS

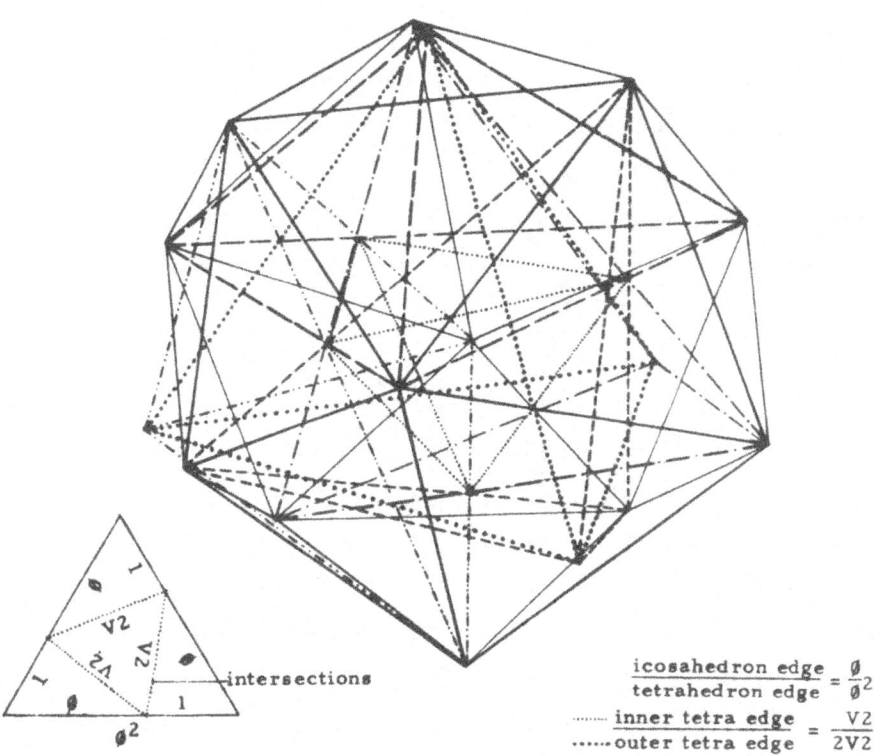

$$\frac{\text{icosahedron edge}}{\text{tetrahedron edge}} = \frac{\emptyset}{\emptyset^2}$$

$$\frac{\text{inner tetra edge}}{\text{outer tetra edge}} = \frac{\sqrt{2}}{2\sqrt{2}}$$

icosahedron formed by 4 positions of tetrahedron in rotation

ROTATION IN THE HIGHER SOLIDS, continued

<u>icosahedron formed by 4 positions of tetrahedron in rotation--</u>
base planes of 4 tetrahedrons intersect in the same way as
the 4 triangles below, with edges divided in the Divine Proportion. 'vestigial' polarity is found in the intersections
forming a small inner tetrahedron and in the extremities
forming a large outer tetrahedron, each with point to face
polarity with the other.

<u>icosahedron formed by 4 positions of equil. triangle in</u>
<u>rotation</u>--icosahedron also formed by 3 Golden Rectangles

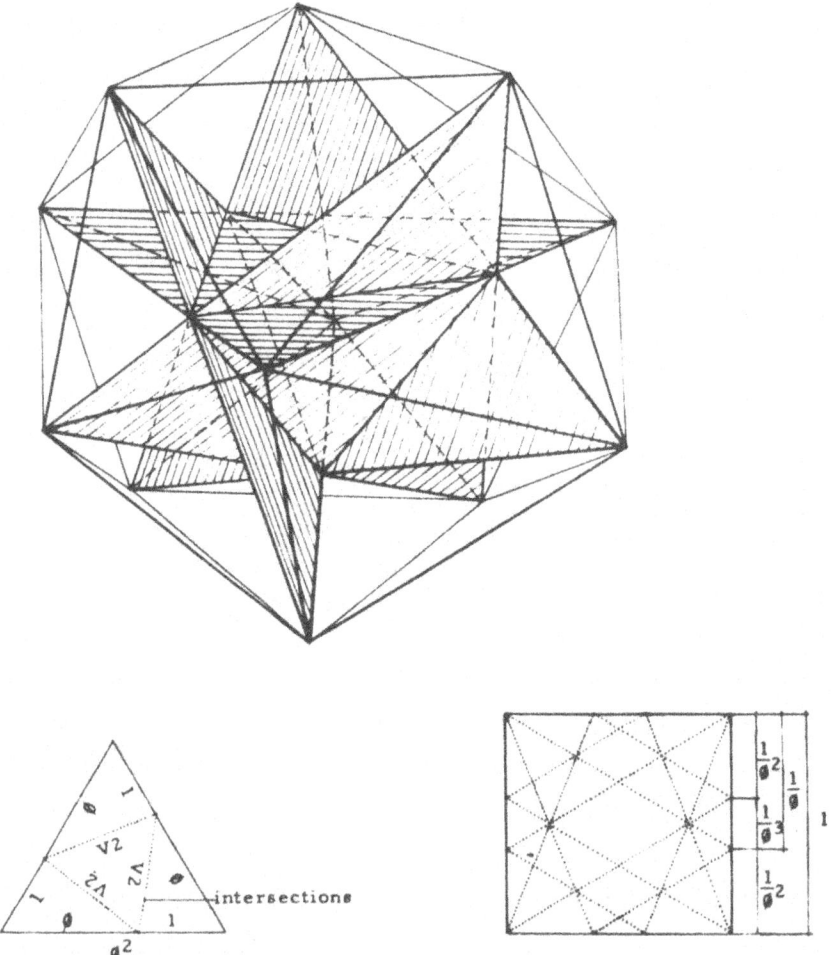

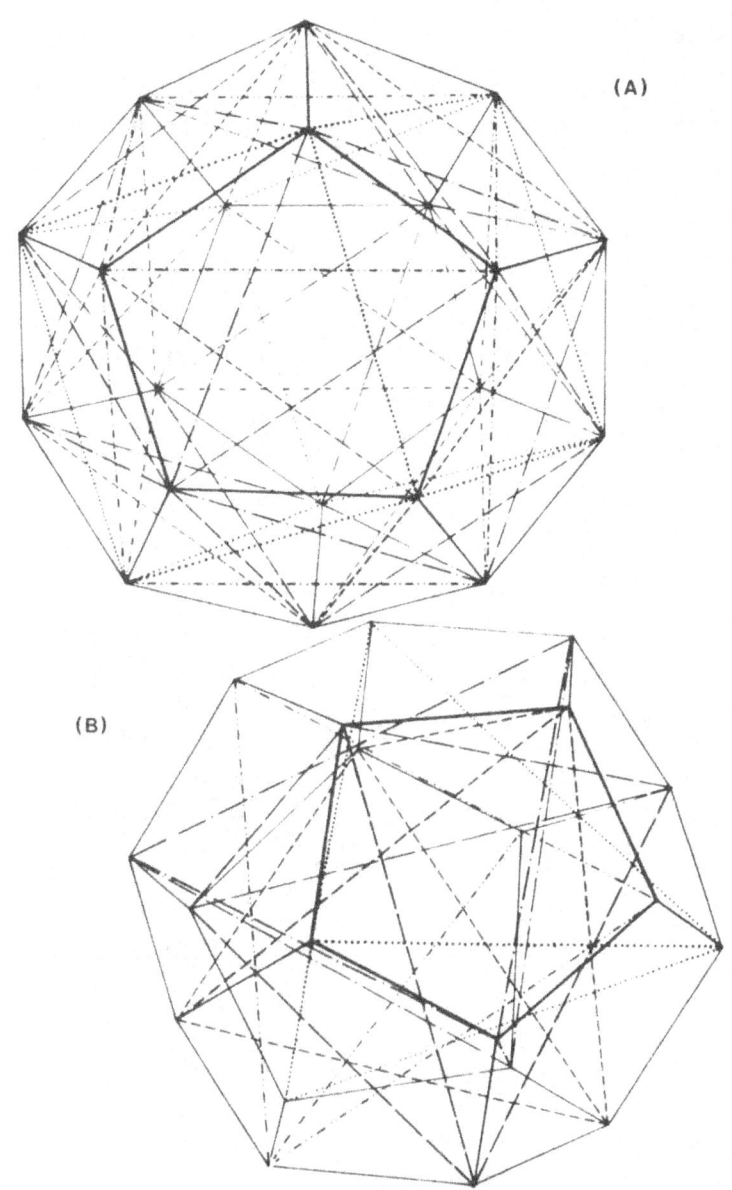

(A) : <u>dodecahedron formed by 5 positions of cube--</u>
in rotation-intersections divide all cube edges in Divine
Proportion-<u>edge of cube</u> $\dfrac{1}{1/\emptyset}$
       edge of dodecahedron

(B) : <u>dodecahedron formed by 5 positions of tetra-</u>
<u>hedron</u>--in rotation-intersections divide all tetrahedron edges
in Divine Proportion-<u>edge of tetrahedron</u> $=\dfrac{\sqrt{2}}{1/\emptyset}$
       edge of dodecahedron

ROTATION IN THE HIGHER SOLIDS, continued

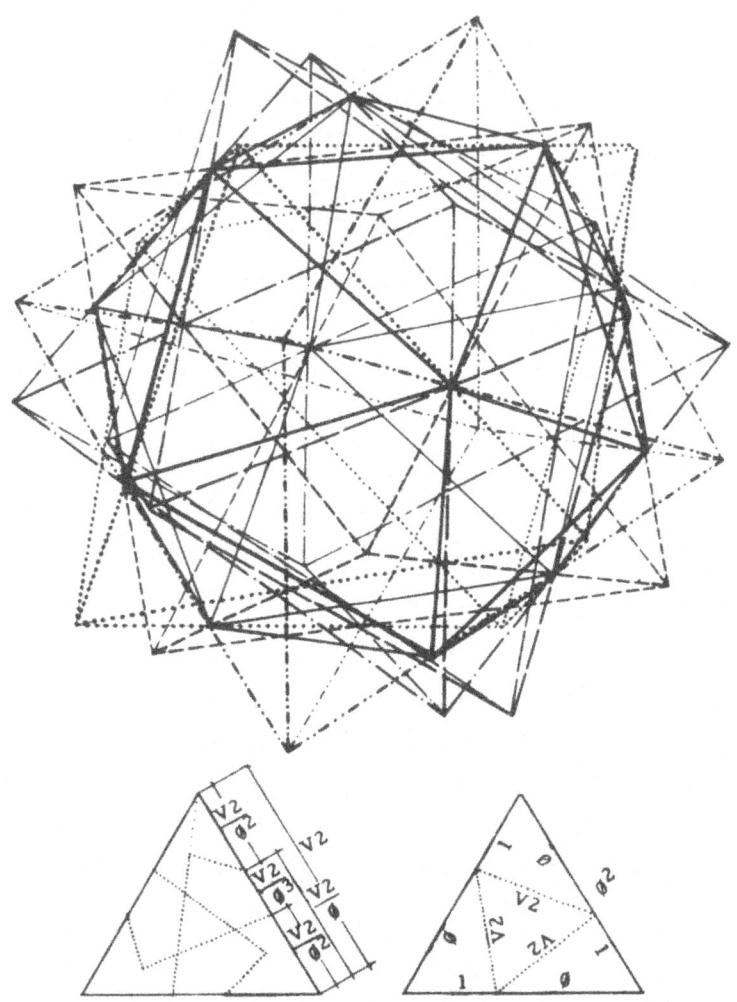

icosahedron formed by 5 positions of octahedron in rotation --
icosahedron vertices occur at Divine Proportion intersections
of octahedron edges-octahedron vertices form icosidodecahedron --
$$\frac{\text{edge of octahedron}}{\text{edge of icosahedron}} = \frac{\emptyset^2}{\sqrt{2}}$$

This next order of *rotational* symmetry, abstracted as a circle or sphere, is based on 5 or 10 fold symmetry around a point. The semi-circular arch, vault and dome are architectural expressions of this order. The prehistoric circular 'sun' temple, or observatory, at Stonehenge, the semi-circular amphitheaters and round tholos temples of the Greeks, the great dome of the Pantheon and the multiple domes of Hagia Sophia are powerful symbols of this geometric ordering principle. Plato described the soul as a sphere, and Silesius is quoted as saying, "God is my center when I close him in; my circumference when I melt in him." (3) Kepler, who determined the law for the ellipticle motion of planets, drew diagrams of the soul in the geometry of five and tenfold symmetry. He wrote,

> "But, although every soul bears within itself a
> certain idea of the circle - a circle not merely
> detached from matter but also somehow from magni-
> tude... so that in this case circle and centre
> almost coincide and the soul itself can be called
> a potential circle as well as a point differenti-

Stonehenge

3Campbell, J. *The Mythic Image*, Princeton 1974, p.64.

ated according to directions and this somehow
qualified–nevertheless there must be observed the
distinction that some faculties of the soul have
to be considered rather as circle, others as
point." (4)

Kepler also proposed a 'nesting' of all five Platonic Solids,
the octahedron within an icosahedron within a dodecahedron
within a tetrahedron within a cube, as the basis of the rotation of the five then known planets, his early perception
of ordering in the universe.

Earlier Copernicus depicted the Universe in sun-
centered concentric rings. At vastly different scales, this
geometric order of circles and spheres is used in current
scientific representations of atoms and of cells.

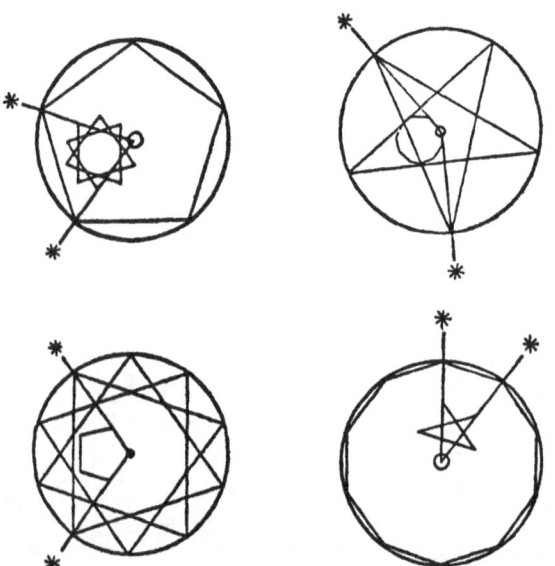

Circumferential figure and central figure

From Kepler's *Harmonices mundi,* Book **IV:** *De configurationibus
harmonicis,* Ch. 5 (Frisch, **V, pp.** 238 and 239: figs. 32, 33, 34, 35).

Kepler's diagrams of the soul

---

[4]Kepler, J. *Harmonices mundi* Book IV, Proposition VI

"The 19th century German chemist Kekule, research-
ing into the molecular structure of benzene,
dreamed of a snake with its tail in its mouth...
He interpreted the dream to mean that the structure was a closed carbon ring." (5)

Circular symbols were used by the Egyptians and Aztecs in
representing a sun god, and in the orient divine birth occurs from the circular form of the lotus flower. The circular symbol of the wheel occurs in the Tibetan Wheel of
Life and in the Tarot as the Wheel of Fortune. The psychologist Carl Jung wrote, "We know from experience that the
protecting circle, the *mandala*, is the traditional antidote
for chaotic states of mind." (6) The motif of the protecting

Tibetan Wheel of Life

5Jung, C.G., "Approaching the Unconsciousness", *Man and
His Symbols* edit. C.G. Jung Doubleday 1964 p.38

6Jung, C.G., Collected Works, vol.9:1 *Archetypes of the
Collective Unconscious* Pantheon 1959 p.10

circle or space, the 'temenos' or magic circle, occurs in
Medieval towns as the fortified ring-wall. This protective
wall is elaborated in the Renaissance designs for ideal
cities, which use angled fortified walls to surround the
city with star-like geometries, many with five and ten fold
symmetry. Such a *rotational* symmetry includes within it the
bilateral symmetry of the street system, an intuitive use of
the basic geometric sequence, which includes *bilateral* within *rotational* symmetry.

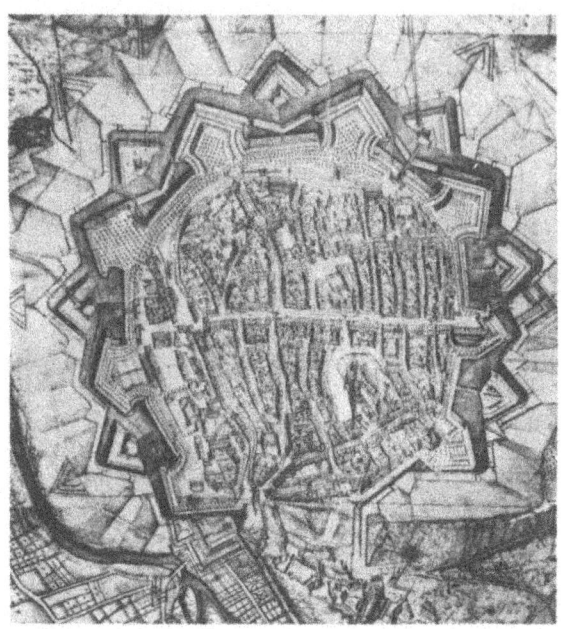

Freiberg, Germany

The 'fourth dimensional' extensions of these *rotational* forms along an axis perpendicular to the radius of rotation, expressing again the *tension of polarity,* define the
helical forms of the geometric progression. Since both the
rotational forms have pentagonal symmetry around a center,
the plan of their *helical* extensions is based on the (10
sided) decagon with its side in Divine Proportion to its
'radius' (of the circumscribed circle). The vertical extension of each turn is in Divine Proportion ratio to the side
of the decagon, making a Divine Proportion progression -
vertical turn = $\emptyset$, horizontal turn = $\emptyset^2$, and radius of
turn = $\emptyset^3$.

D N A

H E L I C A L   E X T E N S I O N   O F   R O T A T I O N

helical extension of dodecahedron
(or icosahedron) along the axis
perpendicular to axis of rotation
can define double helix similar to
the structure of D N A molecule

plan below of even numbered turns

rotating dodecahedron forms decagon
with 10 turns/circumference-in each
turn a Divine Proportion progression-
vertical increment of turn ϕ
horizontal increment of turn ϕ^2
radius of turn ϕ^3

plan below of odd numbered turns

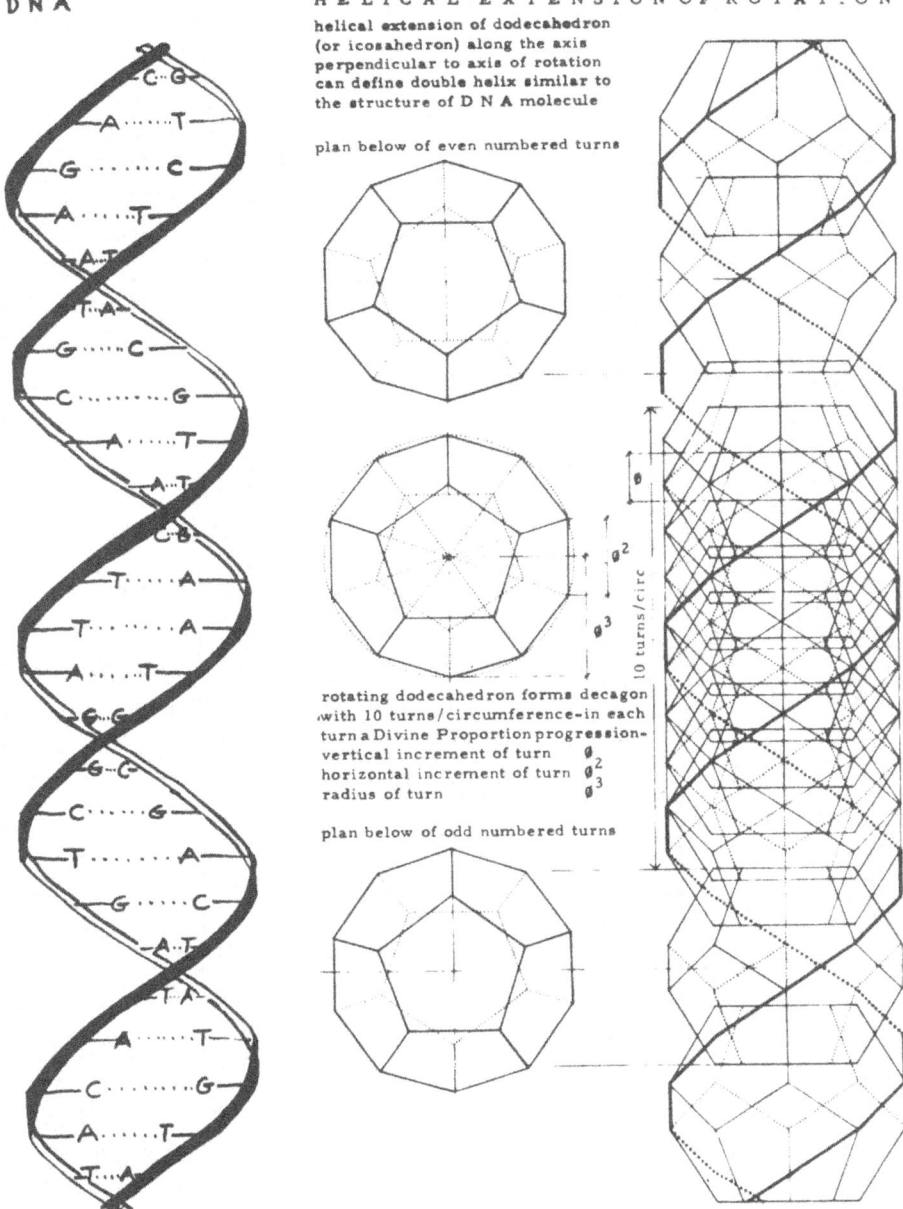

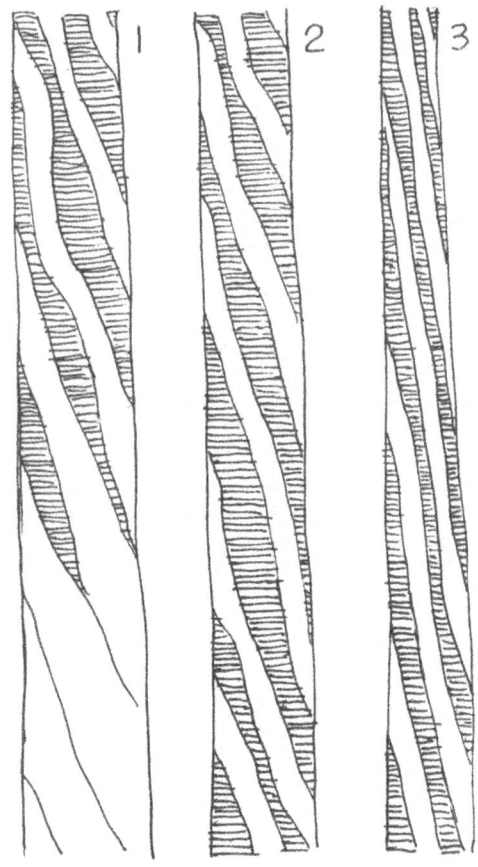

Helianthus annus

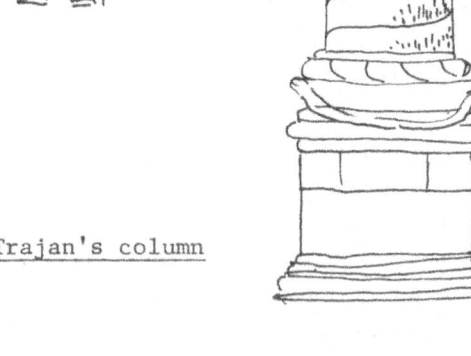

Trajan's column

The ordering principle of the *helix* is less easily recognized by its helical geometry than it is by its verticality, since it appears frequently as a simple cylinder. If we look closely, however, we can see, in natural forms such as the stem of the *Helianthus annus*, a surface helical pattern, which in this case predicts the position for potential growth of leaves. Similarly, Trajan's memorial column in Rome is embellished with a helical pattern from bottom to top illustrating a history of the Dacian wars. Many such columns in churches are decorated with a similar helical design depicting the life of Christ. In other examples of this vertical order, such as the primitive phallic form of the Menhir of Brittany and the more sophisticated form of the Egyptian obelisk, there is no helical articulation. Yet the universality of these isolated symbolic column forms and the inclusion of the helical serpent on the Greek Herm suggests their relation to this form empathy. In the Greek Herm, we see the symbol of entwined serpents combined with the phallus as a form which was placed at crossroads "as leader of souls to and from the underworld" (7) Referred to

7Jung, C.G., "Approaching the Unconscious", *Man and His Symbols*, p.156

as the *lingam*, "the phallus functions as an all-embracing
symbol in the Hindu religion" (8) and is monumentalized in
stone along the roadside. As a symbol the helical serpent
emphasizes the geometric polarity of the helix form. Often
pictured coiled on a cross or tree, the serpent symbolizes

Medieval manuscript depicting polarity of helical serpent

the polarity of both a savior and a devil, both a cthonic instinctual and 'most spiritual' animal. The serpent is depicted entwined on the tree of knowledge of good and evil by
Michelangelo in his Sistine Chapel painting of the Expulsion
of Adam and Eve from the Garden of Eden. As the connector
between instinct and spirit, the serpent is a symbol of healing and it is universally the emblem or badge of the medical
profession. In architecture, the building of towers and sky-
scrapers such as the twin Marina towers in Chicago, London's
Post Office tower and the Tokyo tower, expresses this vertical
form empathy. The geometry of circular towers occurs in
abundance in Medieval churches and castles. Where they are
added to the earlier ring-walls of towns as fortifications,
they intuitively articulate the basic geometric sequence of
bilateral to *rotational* to *helical,* as *helical* towers around
the *rotational* wall enclosing a *bilateral* street pattern.
More striking helical forms in architecture are the exterior
helical stairway of the Francis I wing at the Chateau de
Blois, probably inspired by Leonardo, and the helical stair
of the Palazzo Contarini in Venice. At Chambord, in addition to a number of exterior helical stairs, the interior
central stair is a double helix, or two intertwining stairs
which never meet. Again, it was Leonardo, whose sketch of a
double helix stair inspired the Chambord stairs. It is particularly fascinating, in the light of Leonardo's profound
interest and research in natural forms and human anatomy,
that his intuitive leap into the geometry of the double
helix prefigures the discovery of the double helix structure
of the DNA molecule in 1953 by Watson, Crick, Franklin and
Wilkins, and the earlier discovery of the single alpha helix
by Linus Pauling.

The *spiral,* the most complex order of the geometric sequence, expands the cylindrical helix horizontally at its
base to form a cone-shaped spiral. It is often abstracted
as a flat spiral. A proportional increase in the radius of
rotation of the *helical* forms, expressing *rotational tension*
results in *spiral* forms, the fourth stage of complexity in
the cycle. The only ratio which satisfies the condition of
a logarithmic spiral in which width of turns increases at a
fixed ratio to length is again the ratio of the Divine Proportion. The shifting order of these forms between polarity
and rotation includes the previous order within the new
order, so that *rotational* includes the polarity of *bilateral,*
helical with its own polarity includes rotation plus polarity, and *spiral* with its own dominance of rotation includes
polarity plus rotation, plus polarity. From *bilateral* to
rotational to *helix* to *spiral,* these geometric examples of
polarity and rotation provide us with precise examples in
the formative process from simplicity to complexity in the

Left. Palazzo Contarini, Venice
Right. Albi Cathedral drawn by Louis I. Kahn

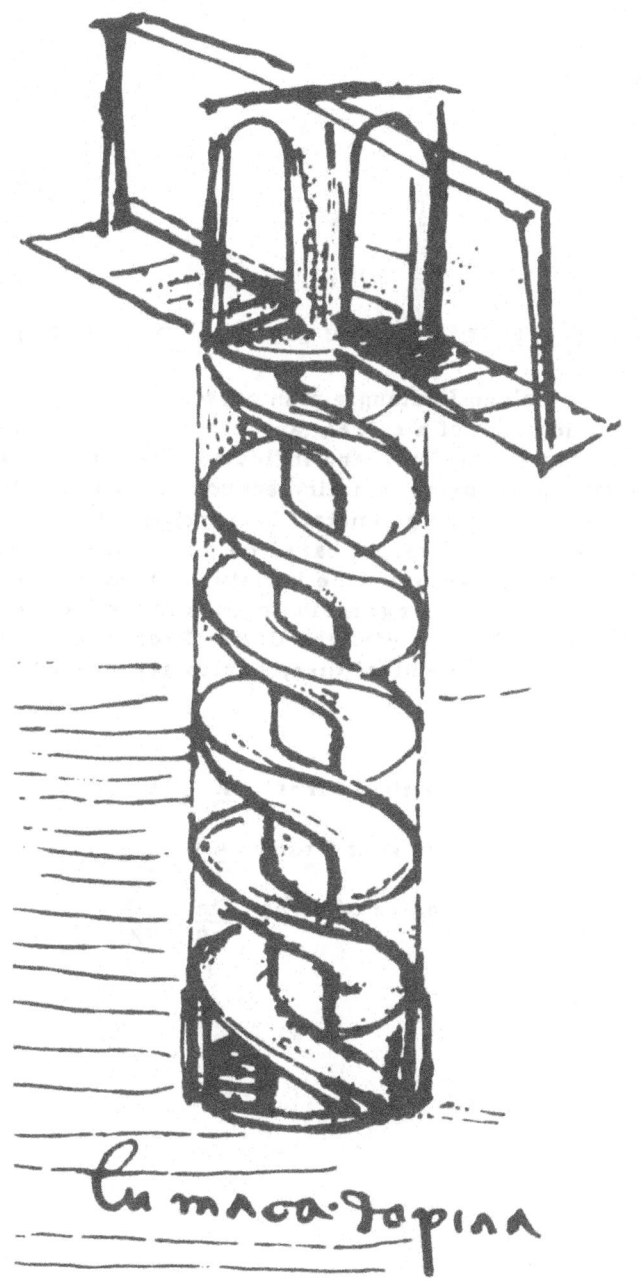

Leonardo da Vinci's double helix stair

S P I R A L   E X T E N S I O N   O F   H E L I X

using the Golden Rectangle plan of 'whirling squares' as
geometric basis of a spiral, any of the 5 solids may be
substituted for the cube and indicated with a complexity
of overlapping forms to articulate polarity within rotation
within polarity- the geometric hierarchies of spiral form.
other proportional series based on the Divine Proportion
can be used to define a wide variety of spiral forms. the
Divine Proportion progression occurs in the increase or
decrease in size of units- the Divine Proportion relations
in dimensions of vertical turn, horizontal turn and radius.

$$\text{vertical turns} \quad \frac{X}{Y} = \frac{Y}{Z} = \emptyset$$

$$\text{horizontal turns} \quad \frac{AB}{BC} = \frac{BC}{CD} = \emptyset$$

$$\text{radius of turns} \quad \frac{OA}{OB} = \frac{OB}{OC} = \emptyset$$

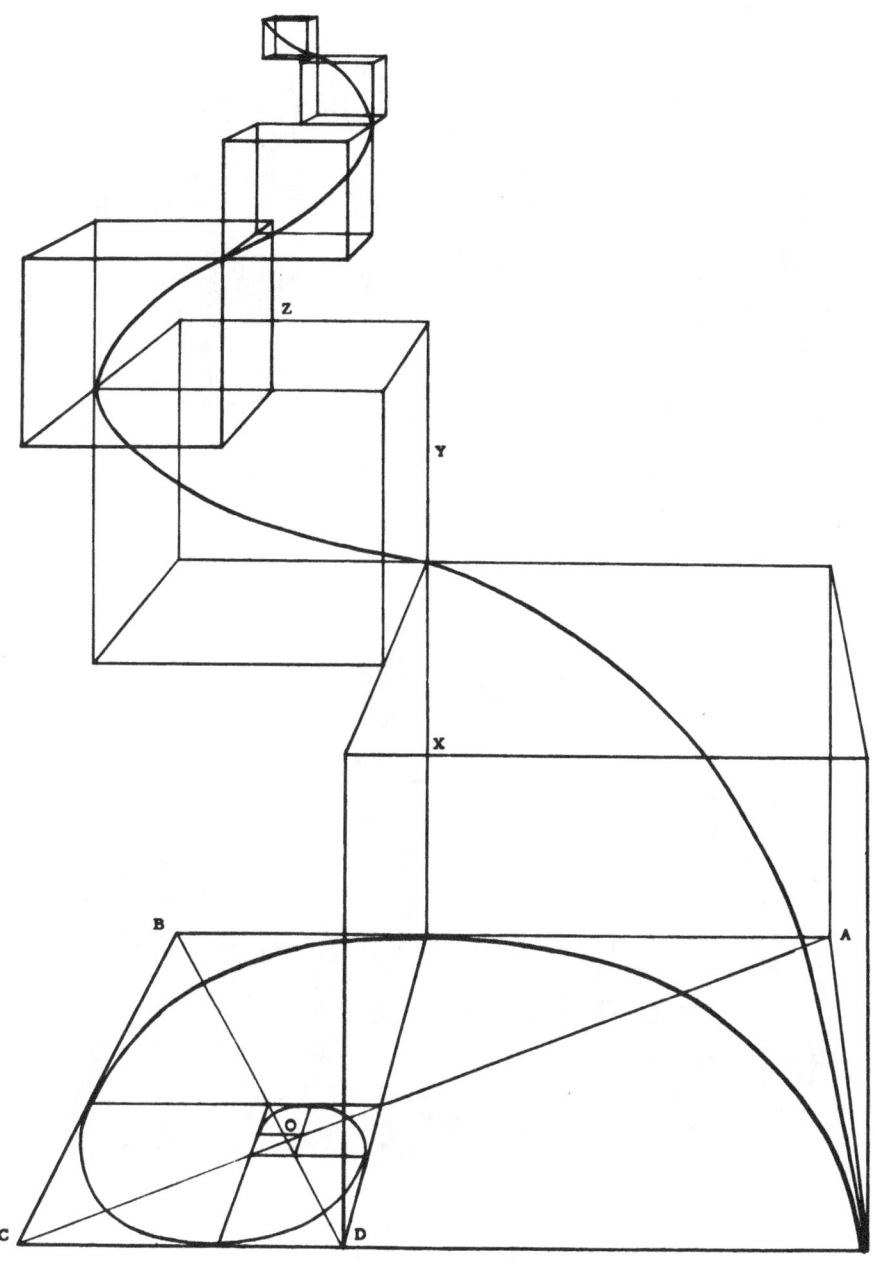

Top. Frank Lloyd Wright's Guggenheim Museum
Bottom. "Tower of Babel" painted in 1563 by Peter Bruegel.
     Reproduced by permission from the original in Kuns-
     thistorisches Museum, Vienna. All rights reserved.

gradual intensification of structure by new extensions in
space, leading from the rigid incompressibility of the *bilateral* tetrahedron to dynamic *rotation* to flexible flow of
helix and to coiled resiliency of the *spiral*.

The ordering geometry of the flat spiral finds expression in the tortuous maze of the Cretan labyrinth as well as
in the spiral ritual path of healing of the Navaho Indians.
In gothic cathedrals, the floor patterns of processional
mazes can be walked as symbolical pilgrimmage to the Holy
Land. The cone-shaped spiral occurs in primitive mountain
forms, and is represented by Dante in his *Divine Comedy* to
express progressive stages of human development. Breughel
the elder's painting of the Tower of Babel shows a similar
empathy for the spiral geometric order. Jung writes of the
spiral,

> "It is the very symbol of unfolding...in plants
> the buds or the beginnings of leaves are arranged
> in a spiral...it is the functioning of opposites,
> the reconciliation of opposites. The man who dis-
> covered the mathematical law of the spiral is
> buried in my home town of Basel; on his tomb-
> stone, the Latin inscription can be literally
> translated, 'In an identical way, changed, I lift
> myself up.sameness and non-sameness..the spiral
> is a very apt symbol to express development." (9)

This spiral geometry does, in fact, operate in the laws of
phyllotaxis, which govern the arrangement of leaves around
a stem. Similar spiral patterns occur in the florets of the
sunflower, the daisy, in the pine cone and in the surface
patterning of the pineapple. As in the geometry of the
Platonic Solids and the helix, these interlocking spiral
patterns embody incremental relations in the Fibonacci sum-
mation series (this series 1, 1, 2, 3, 5, 8, 13, 21, 34, 55,
89, 144, 233,.., in which each number is the sum of the two
preceding numbers, by its twelfth number ((144)) becomes a
logarithmic series. It also very quickly developes a Divine
Proportion ratio ((1.61803...)) between successive terms
((233/144 = 1.61805...)) . I have proposed elsewhere (10)

---

[9]Jung, C.G., "The Interpretation of Visions, Excerpts
from the Notes of Mary Foote", *Spring*, The Analytical Psychology Club of N.Y. Inc. 1962 pp.135-7

[10]Tyng, A.G., *Simultaneous Randomness and Order: the
Fibonacci Divine Proportion as a Universal Forming Principle*
PhD diss. 1975 University Microfilms International, Ann
Arbor Michigan, London England

Spiral Galaxy

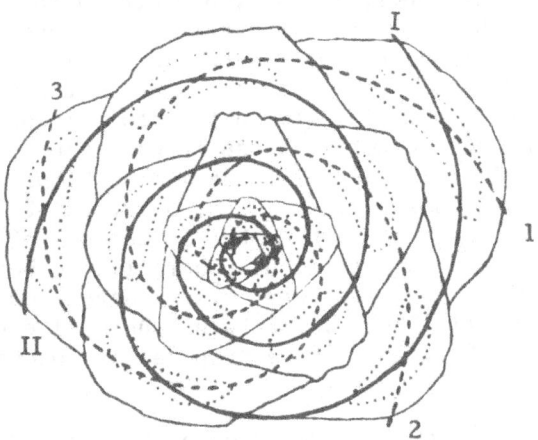

Quinqueloculina Seminulum 2 + 3
(after A. H. Church)
Relation of Phyllotaxis to Mechanical Laws
American Botanical Memoirs XV, 1901

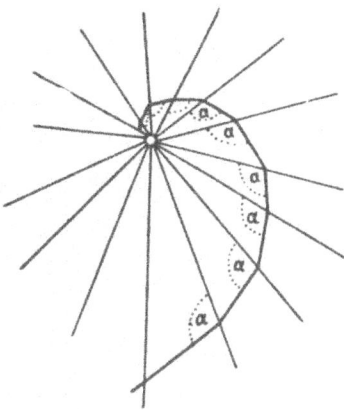

Spiral path of an insect, as it draws toward a light.
From Wigglesworth (after van Buddenbroek).

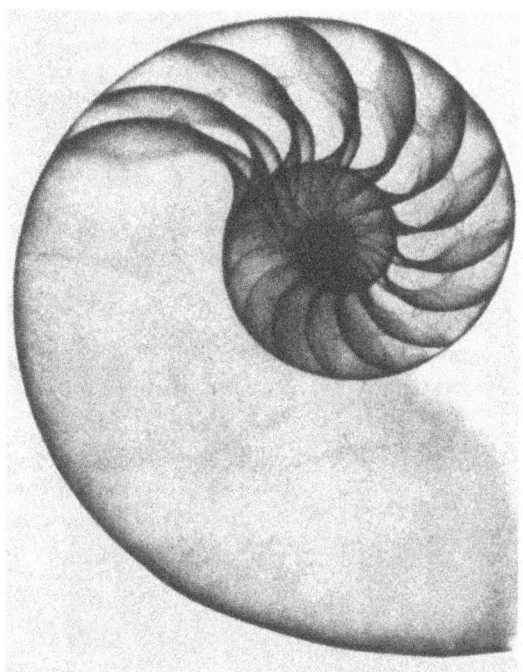

Nautilus Shell

Top. Boromini's S. Ivo
Bottom. Ziggurat of Samarra

that it is *The Probability Mean* and that as a Fibonacci-
Divine Proportion matrix it operates as a universal forming
principle for light and matter.) In addition to the occur-
rence of this spiralling Fibonacci order in plants in the
laws of phyllotaxis, this spiral order governs the shell
structure of snails, the shell of the *nautilus* displaying an
elegantly precise geometry. At vastly different scales, we
see this order in the Lorentz force, which tends to make
charged particles of electrons and protons take a spiral
path through the earth's magnetic field, we see it in the
spiral path of an insect as it flies toward the light (11)
and we see it in the movement of stars and planets which
gives spiral form to galaxies.

In architecture, a pure expression of this form empathy
is the 9th century ziggurat of Samarra, the convergence of
its spiralling ramp reaching toward an infinite point of
synthesis. This geometric order occurs in more abstract
form in tapered Gothic spires and stepped flying buttresses.
Boromini's lantern of S.Ivo and Guarini's tapering lantern
of San Sindone restate this ordering principle in energized
rococo geometry. The French symbol of the Eiffel Tower
pioneered new structural principles in its articulation of
this form empathy, while Gaudi's Sagrada Familia finds a
different articulation of the same order in its perforated
tapered towers. Frank Lloyd Wright found physical forms to
express this geometry in his proposed Mile High Tower and in
the geometric inversion of his Guggenheim Museum.

This geometric sequence of *bilateral, rotational,
helical* and *spiral* not only occurs as a logical building up
of complexity of geometric order, but, in the history of
architecture, it has occurred as a *repeating* sequence of
underlying form empathy, which manifests itself in recurring
cycles, each somewhat different in articulation or 'style'.
I have traced 11 such cycles from the time of the Great
Pyramid of Gizeh to the present (12). One such cycle began
in Italy with the Renaissance, which indicates in its name
the sense of rebirth, its four phases marked by shifting empathy from the forms of Proto-renaissance to High
Renaissance to Baroque to Rococo. This ordering geometric
sequence shifts from the simple, serene, symmetrical, horizontal, rectilinear forms of the palaces of the merchant
princes, Strozzi and Ruccelai, to more rounded open forms of

---

[11]Thompson, D'Arcy W., *Growth and Form*, Cambridge
(Eng.) 1952, p. 756

[12]Tyng, A.G., "Geometric Extensions of Consciousness"
Zodiac 19 Milan, Italy 1969, pp.153-160.

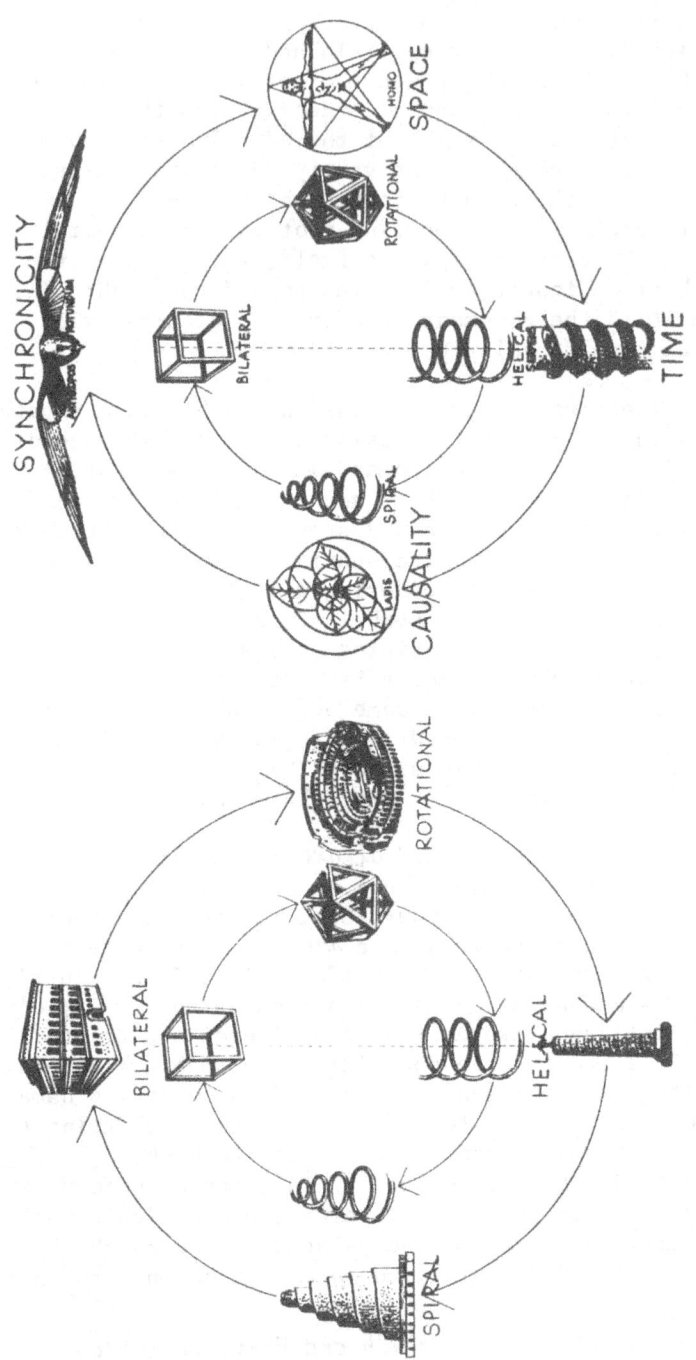

Left. Architectural cycle of shifting form empathy

Right. Psychic cycle of shifting archetypal images

domes and arches, the dome of St. Peter's in Rome and of the
Duomo in Florence, to the helical stairs at the Palazzo
Contarini and those inspired by Leonardo at the Chateau de
Blois and Chambord, as well as the monumental verticality of
the obelisks of Pope Sixtus V, shifting again to the spiral-
ling undulation of facades and lanterns of Boromini and
Guarini.

The biological roots of shifting human empathy may be
found in the most fundamental living forms. While the *repeating* cycle of *bilateral, rotational, helical* and *spiral*
apparently is not valid for nonliving or 'inorganic' forms,
it clearly appears to be a special achievement of living
forms. The energies and configurations *progressively* built
up in the rhythmic interplay of rotation and polarity result
in the *gradual* intensification of structure and the flexible
vitality which is a special achievement of 'higher' living
forms. In natural forms, as in a Renaissance of human consciousness or in a rebirth of the individual psyche, a more
dramatic ecological reconnection or hierarchy is formed when
the ultimate complexity of the 'spiral' forms is included in
a new simplicity of 'bilateral' form at larger scale. The
structure of hemoglobin, the molecule which gives blood its
red color, is a beautiful example of such a hierarchy. It
includes 'bilateral' tetrahedron carbon atom bonds and octahedral iron atom bonds, it includes 'rotational' pentagonal
ring nitrogen bonds, it includes 'helical' alpha and beta
helices, and 'spiral' coiled coils, or irregularly spiralling helices, around each heme molecule. Yet its total form
is ordered by the simple 'bilateral' positioning of the four
hemes at the four corners of a larger tetrahedron, its complete form recalling the basic carbon atom's tetrahedral bond,
the single overall order having affinity with the multiple
inner order. This reconnection of symmetry is again an
ecological as well as hierarchical principle. Hemoglobin
includes forms which increase in scale and complexity along
with identifying shifts of symmetry, from 'bilateral' to
'rotational' to 'helical' to 'spiral' with the inclusion of
spiral in a larger 'bilateral' simplicity. In addition to
the interlocking connectivity of these symmetries, and the
inclusive overlapping of simplicity and complexity, there is
also a simultaneous interlocking of randomness and order.
While the 1 to 1 symmetry of 'bilateral' form may offer an
initial *inner* 'head or tail' *randomness* bound by a simply
ordered external form, the most *complex* or 'spiral' level of
form, while *internally ordered* or synchronized through precise Divine Proportion linkage, offers in the *external* form
a more *complex randomness* in its possibilities for new external connections. In other words, the *simplest* form has
internal randomness and *external order*, the *complex* form has

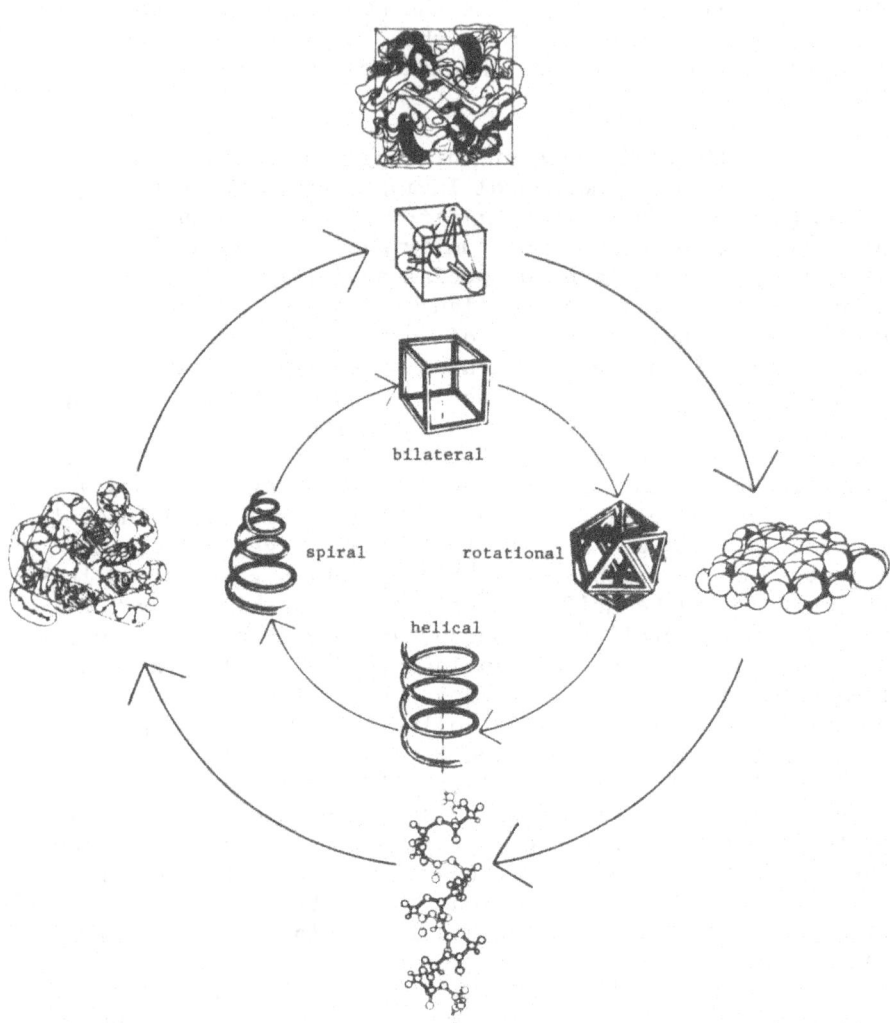

Structure of hemoglobin

inner synchronized *order, external randomness.* Another, rather unexpected, confirmation of the Fibonacci-Divine Proportion matrix occurs in Deoxyhemoglobin, one of the two forms of hemoglobin resulting from the relative motion of its alpha and beta helices. In its four hemes, I have found that, if the 24.7 angstrom unit distance between hemes is taken as a value of 1, then its 34.9 angstrom unit distance will have the value of the $\sqrt{2}$ or 1.414, while the 39.9 angstrom unit distance will have the value of 1.618 or the Divine Proportion. Thus, its total tetrahedron form, reconnecting to carbon atom bond forms, also reconnects, in the relative distance between its hemes, to the Fibonacci ratios of its carbon atom bonds and to the Divine Proportion ratios of its helices. Thus hemoglobin's total ordering of its 100,000 atoms binds its identity with the Fibonacci-Divine Proportion matrix and establishes a clearly recognizable hierarchy of form.

With tremendous leaps in the scale and complexity of life forms, when the geometry is camouflaged by variations in color, motion and mysterious habitats, obscured by lapses in time and hidden embryo shelters, the cycles of symmetries are less sharply defined. The life cycle of the butterfly is clearly defined in four phases: the *rotational* symmetry of its eggs, the *helical* symmetry in its form as a caterpillar or larva, the *spiral* symmetry of the pupa or chrysalis form and its dramatic rebirth in a magnificent form of *bilateral* symmetry. The frog follows a cycle from the *rotational* symmetry of the zygote, to *helical* embryonic body-stalk, to *spiral* form of the tapering tadpole to *bilateral* symmetry of the mature frog. *Bilateral* man, evolved from numberless hierarchies of cycles of form, from the primordial ordering of atoms and molecules, goes through the cycle again in the early stages of embryonic development from the *bilateral,* then *rotational* cleavages of the ovum, to the *helical* body-stalk of 18 or 19 days, to the *spiral* embryo of about 4 weeks to the miniature complexity integrated into his ultimate *bilateral* form as a 10 week 2 inch embryo of potential human being.

Julian Huxley has observed that, "the cells orient themselves along the lines of tension, and multiply faster here than elsewhere" (13). Acted on in turn by tensions related to the earth's rotation and to gravity, the pure geometry of atoms and molecules, through the rhythmic interplay of rotation and polarity, has been continuously adding to itself in self-transforming patterns, arranging and rearranging itself in infinite possibilities of form.

13Huxley, J., *Evolution in Action,* Signet Science Library, 1953, p.35.

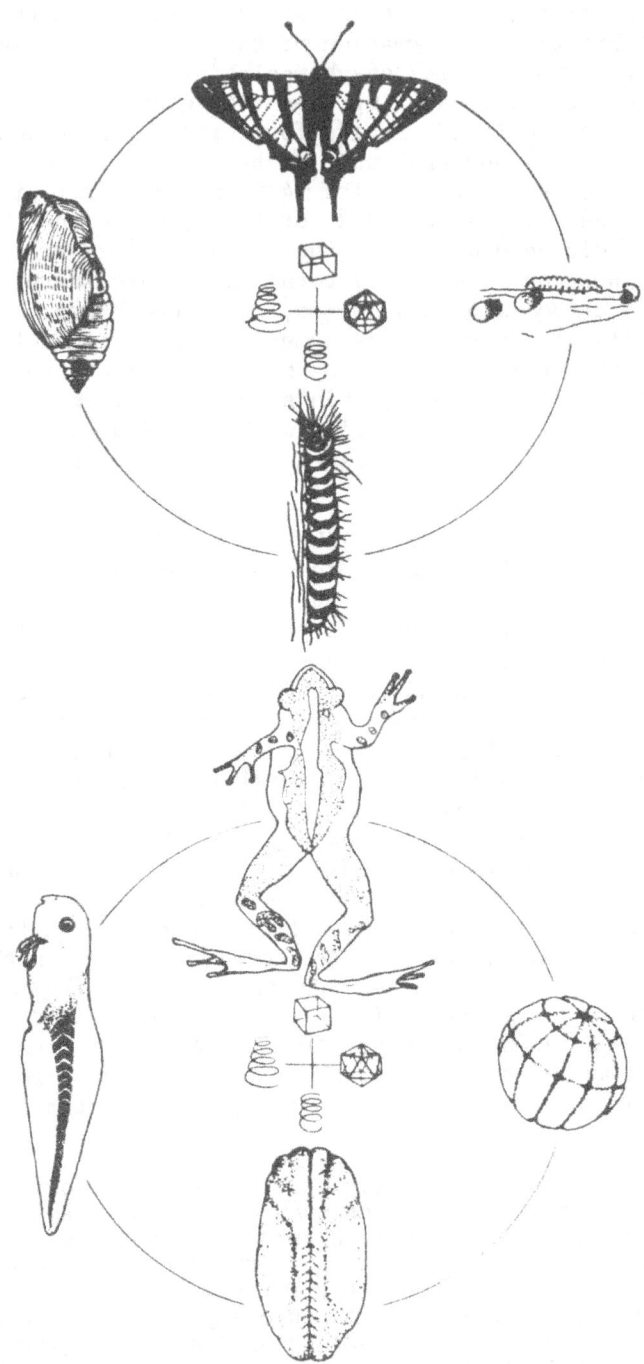

Top: butterfly cycle; bottom: frog cycle

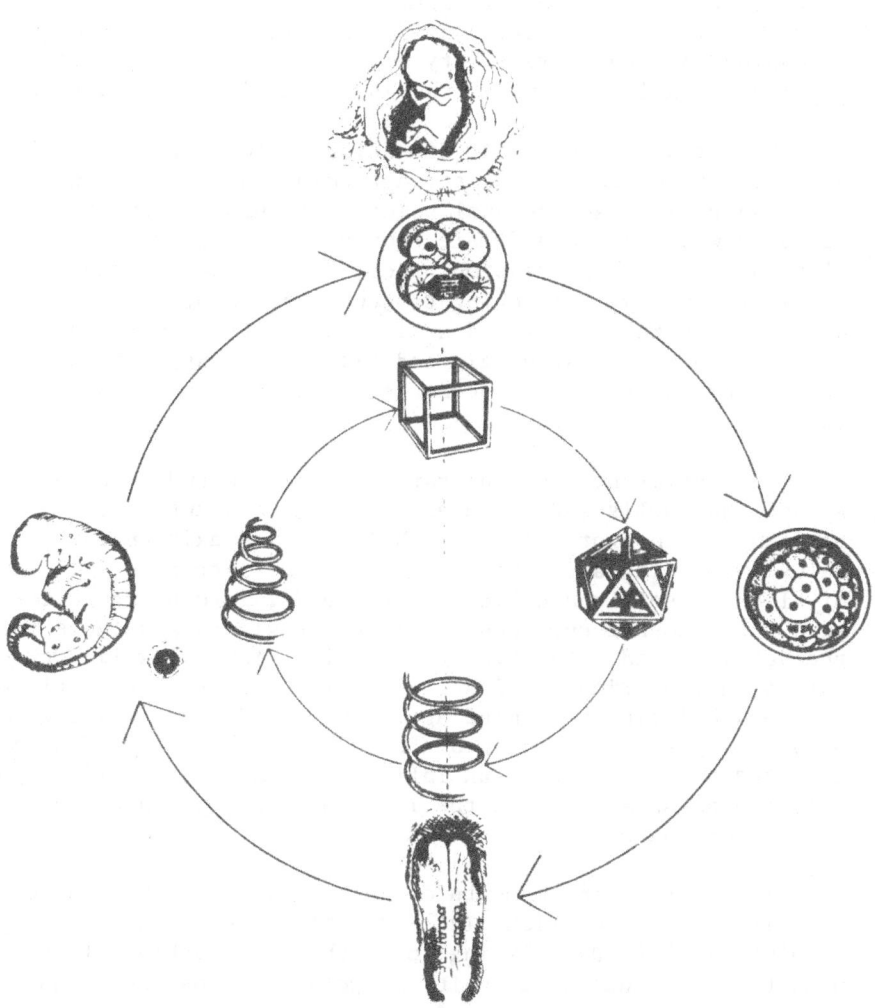

Cycle of human embryo

Form thus finds its own form, extending feelers, gills and tentacles to the world around it, in its *rotational* tensioning, expanding its magic circle to new concepts of *space* -- from deep sea creatures' first sensitivity to light to man's skyward extension of sight through radar telescope, from the first articulation of fin or finger to the spiritual dimensions of man's creativity -- response to the challenging tension of individual man with his collective environment.

Form finds new *helical* dimensions, elongating to differentiate intake and output, strengthening backbone between tusk and tail, head and anal poles, articulating the tensions between male and female from simple reproduction to elaborate courtship, tensioning emotions between sexuality and spirituality, stretching to new concepts of *time* between past and future, memory and anticipation, between the awareness of man's darkest origin and his highest aspiration, between the depths of the unconscious mind and conscious thought.

Form stretches to elaborate both length and breadth in *spiralling* shells and branches, antennae and antlers, experimenting in exotic forms with 'displayed existential value', dividing and subdividing into the intricate filigree of blood vessels and delicate nerve ends, involuting to extend the internal surfaces of digestive glands and lungs for the detailed and intensive organization of complex life processes, discovering for itself infinite variety and complexity -- subtleties of camouflage, heightened movement, the play of light and coloring of forms, *the tensioning of forms in space and time* toward an infinity of matter, toward weightlessness and toward the intricate involution and complexity of the brain.

In the fleeting moments of balance between the tension of polarity and rotation, *the tensions of space and time are resolved* in *bilateral* living form, the interlocking of complexity to produce a new unity of relationships the transformation of the end of complexity to a new beginning of simplicity -- the inclusion of complexity within simplicity -- *the discovery of the cycle* -- the balancing and neutralizing of tensions of space and time within a higher order -- *the creation of the first hierarchy of form.* From countless levels of such hierarchies the brain of man was formed, the evolution of human consciousness and the psychic potentials of 'individuation' and rebirth, man's search for the secret of creation, for concepts of immortality free of time, space,

Hemoglobin

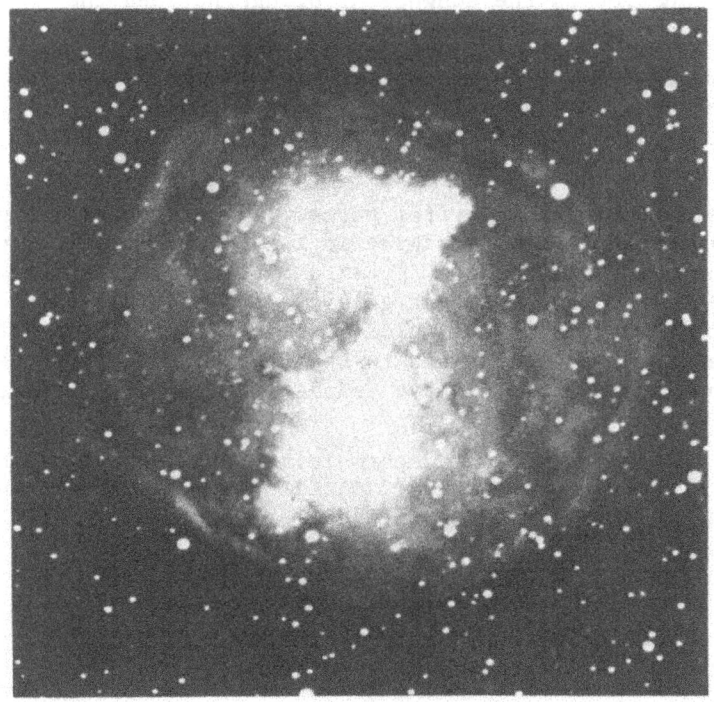

Dumbell Nebula

causality -- for *synchronicity* (14), for the immortal 'static' synthesis of mortal 'kinetic' concepts.

Herbert Read wrote,

> "the artistic activity might therefore be des-
> cribed as a crystalization, from the amorphous
> realm of feeling, of forms that are significant
> or symbolic... Physical imprints or perceptual
> memory have been called 'engrams'; and if we may
> assume that man inherits a physical disposition
> towards images that conform to these patterns,
> then we arrive at Jung's conception of the
> *archetype*, a term in his psychology which indi-
> cated an inherited structure of the brain pre-
> disposing the human race, at certain epochs, to
> the invention of particular kinds of symbol or
> to the creation of particular mythical figures"
> (15).

The Swiss scientist Pauli wrote of them,

> "As ordering operators and image-formers in this
> world of symbolic images, the archetypes thus
> function as the sought-for bridge between the
> sense perceptions and the ideas" (16)

Elsewhere (17), I have indicated the extraordinary correspondence of this geometric ordering sequence of four phases
with the shifting archetypal images of the four phases of
Jung's cycle of 'individuation'.

The human capacity for perceiving and conceiving order
continuously finds its source in the geometric ordering sequence of *square, circle, helix* and *spiral*. At extremes of
scale between molecule and galaxy, the invisible fourfold
order of hemoglobin is linked through a fourfold *mandala* or
archetype of consciousness to the fourfold order of the
'Dumbell Nebula' in distant space. *We 'see' the order.*

14See Jung, C.G., Synchronicity -- An Acausal Connection Principle. In *The Interpretation of Nature and the
Pysche*, with W. Pauli, Bollingen 1955, Patheon.

15Read, H., *Ibid.*, p. 31.

16Pauli, W., The Influence of Archetypal Image on the
Scientific Theories of Kepler. In *The Interpretation of
Nature and the Psyche* with C. G. Jung.

17Tyng, A.G., "Geometric Extensions of Consciousness",
Zodiac 19, Milan Italy 1969, p.146-152

6

Visual Comprehension of n-Dimensions

David W. Brisson

Note to the Reader

This paper is intended to be read in conjunction with the presentation of a set of slides projected on a special screen and viewed by an audience equipped with special polaroid glasses. The images are hyperstereograms and are included at the end of this paper. They may be viewed like ordinary stereograms by "crossing one's eyes." However, they are unlike ordinary stereograms in that they have vertical as well as horizontal parallax. Therefore they may not be "fused" in their entirety at any one moment, but must be scanned by "cocking" one's head. This is very difficult for many persons to accomplish. For a small number it is impossible. It is the conviction of the author that the result is well worth the effort, for the experience, if accomplished, transcends the experience of three dimensions.

-- -- --

The sets of figures that are included in this chapter are stereograms and hyperstereograms of some of the basic hypersolids and complex hypersolids of four or more dimensions. They have two, and in some cases three, dimensions of parallax. To see them properly it would be easier if you could wear polaroid "glasses." What you will be asked to do is quite difficult for the average person and quite impossible for the small number of persons who do not have ordinary binocular vision. You may not be able to experience all of the things that will be discussed. However, even if you cannot, the material offered should present a visual challenge even on a purely monocular level. What you are being asked to do is thus quite difficult. It will require some effort on your part, and the result will be an unfamiliar visual experience. It will be confusing and will take a period of adjustment which should occur slowly as the various figures

are viewed. This brief exposure will only give you a taste
of the experience that is ultimately possible.

The drawings are mathematically precise, constructed
by means of analytic and descriptive geometry. No attempt
will be made to describe them in any great detail. They will
be identified and some of their more interesting properties
discussed, but the purpose of this presentation is to give
some insight into the visual/perceptual meaning of basic
mathematical configurations. For a more detailed discussion
of their properties the viewer is referred to the works of
Coxeter, Manning, Sommerville and the many other
mathematicians who have extensively written on this subject.

The ordinary stereogram will be considered first. In
Figure 1 you will see a stereogram of a cube, which consists
of two different images of a cube, from slightly different
positions. If you have polaroid glasses you will find that
if you alternately blink each eye, each eye "sees" only one
of the two images strongly. Now, touch your nose with your
right index finger, and watch it as you slowly move your
finger away from you and toward the figure. As you do you
will observe that the two bright images of the cube move toward each other until at a certain point they are superimposed.
By this time, many of you will have already "fused" the
images into a single, small, transparent three-dimensional
cube floating in space between you and the page. Unfortunately, for various reasons, a small percentage of you, as has
already been pointed out, will not be able to do this. Don't
give up at this point however, because it takes some people
a long time. Even if you can't do it at all it doesn't mean
that there is something wrong with you. Don't force the
situation, relax a little.

If you let the two images float apart again and
compare them you will note that the difference between them
occurs as a horizontal displacement. All of the corresponding points of the two images are on exactly the same level
vertically, but they hold different relative positions
horizontally. This is accomplished by rotating the cube
around a vertical axis. The displacement is called parallax.

Now, "fuse" the cube again. Concentrate your
attention on the back face, noting at the same time that the
front face splits into two images. Focus now on the front
face and note that now the back face splits into two images.
Notice that this splitting is somewhat amorphous and
unstable. Parts of the cube will literally "vanish." Now,

"fuse" the cube again. Concentrate your attention on the back face, noting at the same time that the front face splits into two images. Focus now on the front face and note that now the back face splits into two images. Notice that this splitting is somewhat amorphous and unstable. Parts of the cube will literally "vanish." Now, practice a little. Move back and forth between the front and back. What you are now doing is scanning a three-dimensional "object" with a dynamic mathematical surface called the horoptor. This surface consists, at any given moment, of the points from the projected "object" that have precisely the same location with respect to the centers of retinae of the two eyes at the particular angle of convergence. As the convergence is modified, i.e., as we rotate our eyes toward or away from each other, the horoptor moves toward or away from us. There can be no question that in this respect, the eyes function as a range-finder. Of course this is not the total story but clearly one of the most fundamental aspects of the perception of three-dimensionality is this range-finder effect. In everyday experience we are usually not conscious of this process. The demonstrable simultaneous suppression of such double images, usually referred to as "retinal rivalry" further conceals from us the mechanics of the process. Most of us, unless we are examining our eyesight clinically, tend to believe that we are seeing one and only one image, while in actuality as you have just observed, we see a highly complex "mosaic" composed of parts of two distinct images that in itself is highly unstable.

Before we leave the cube, there is one further exercise that we should examine. Note that by "cocking" slightly your head from side to side you can vary the apparant vertical relation of the two images with respect to each other. This is possible because the two images are displaced some distance from each other on the page. Practice a bit moving the two images up and down with respect to each other.

Now, in Figure 2 you will see a hyperstereogram of a hypercube or tesseract, the four-dimensional analog of the cube. If you try to fuse this set of images as you did the cube, you will probably experience some confusion, because the two images will not so fuse in their entirety. Since rotation in four dimensions constitutes rotation around a plane rather than around a line as in rotation in three dimensions, the rotational parallax occurs in two directions that are at right angles to each other, and at the same time. This means that the displacement occurs vertically as well as horizontally. The problem here is to somehow utilize this

double parallax to "see" the hypercube in a manner analogous
to that used to "see" the cube.

Using your eyes as you did in the case of the cube,
bring the two images together so that they are superimposed.
Remembering that in the case of the cube you could change the
vertical orientation of the two images with respect to each
other by cocking a little bit your head from side to side,
slowly cock your head from left to right and back, keeping
your head facing directly toward the figure. Some of you
will have already noticed a strange effect. At various
positions of thus tipping your head, different parts of the
image fuse, leaving the rest of the image as double images.
For example, in one position you will see the cubic cell on
the right side as a fully three-dimensional cube, while
tipping your head in the opposite direction the cubic cell of
the hypercube on the opposite side of the hypercube fuses
into a fully three-dimensional cube. While one cube is thus
fused, the other cube splits into a double image. This is
directly analogous to looking at first the front face and
then the back face of a three-dimensional cube, as you did
earlier. In other words, you are scanning a hypercube
binocularly with a three-dimensional horopter.

At this point a few of my readers will have completely
experienced what I am describing. A larger percentage will
have had "flashes" of the experience and a remaining hopeful-
ly small number will have experienced nothing in particular
at all. To those of you who cannot for one reason or another
experience anything that I am writing about, I can only hope
that the description of the experience will be useful in
terms of understanding something that other persons can experience. For those of you who are just beginning to get the
hang of the thing so to speak, I would like to suggest that
you look at the figures for a while longer, before we proceed to the next step.

For the next few figures you are going to be presented
with some of the simpler regular and semi-regular polytopes
of four dimensions. It takes time to fully grasp them, but
perhaps you will get some insight into the diversity, complexity and beauty of some of these figures.

The hyperstereogram in Figure 3 is a truncated hypercube. In
short, the corners of a hypercube have been cut off by three-
dimensional cuts. It is bounded by 8 truncated cubes and 16
tetrahedra.

The hypercube may be further truncated as in the next
figure (4), which is bounded by 9 cuboctahedra and 16 tetrahedra.

It is of cource possible to cut the hypercube in a variety of ways with three-diemnsional cuts. It is cut diagonally into two double-prisms in Figure 5. Note the "ring" of 4 triangular prisms interlocked with a ring of 3 rectangular prisms indicated by the solid lines.

Figure 6 is a <u>lattice</u> of hypercubes, which close-pack to fill four dimensions in the same way that cubes close pack to fill space of three dimensions.

Figure 7 is a 16-celled polytope, one of the Platonic hypersolids, and Figure 8 is that same figure 1/3 truncated.

Figure 9 is a lattice of 16-celled polytopes.

If the 16-celled polytope is fully truncated it forms another Platonic polytope, the 24-cell (Figure 10), which also forms a 4-space filling lattice (Figure 11).

There are many other such polytopes, some quite beautiful and interesting, but unfortunately here we can only give you an overall taste rather than a complete compendium. Next we will consider some curved figures of four dimensions. Figure 12, for example, is a cylindrical double-prism, and Figure 13 is a circle-based hypercone.

A most complex but beautiful form is the right-double-cylinder (Figure 14). It consists of a double prism in which all of the generators are circles.

Next we have two circular sections of isocline planes joined by a ruled surface that exists only in four dimensions (Figure 15). They intersect in only one point, a strange and interesting property of four dimensions.

We will now consider some analogs of the circle. We will consider surfaces of constant positive curvature, and surfaces of constant negative curvature. First we will consider positive curvature.

The two coordinate lines of a plane define the axis of a circle with its center at the origin. Similarly, a sphere may be partially defined by the three great circles of the three coordinate planes of three dimensions as in the stereogram in Figure 16. Analogously, there are the six coordinate planes of 4-space (Figure 17), and there are the great circles of <u>those</u> planes similarly defining a hypersphere (figure 18).

The bounding hypersurface of the hypersphere is curved three-dimensional space. As such, it serves as a model for an elliptic non-Euclidean three-dimensional space.

A curious surface that exists in that hypersurface is the surface of double-revolution (figure 19). It consists of the circles defined in a set of isocline planes as they are rotated around an origin. If the circles are accepted as geodesics of a non-Euclidean hypersurface, then they constitute Clifford's surface, and they have all of the characteristics of the Euclidean plane. Considered as circles in a four-dimensional Euclidean hyperspace, they constitute a curious one-sided closed surface with nearly mystical associations in popular literature: the Klein Bottle.

Now, curved three-dimensional space is interesting, but of even greater interest perhaps, is curved four-dimensional hyperspace, because of the great place that this holds in the General Theory of Relativity.

For that reason we will extend our attention to "spaces" of more than four dimensions, specifically to a curved four-dimensional space imbedded in the four-surface of a five-sphere. In Figure 20, you will see a 5-stereogram of a five-cube, the five-dimensional analog of the cube. I hope by now most of my readers are somewhat used to viewing the hyperstereogram. This 5-stereogram should really not be too difficult to view, for it is little more than a slight modification of the process involved in viewing the hyper-stereogram. As you will probably have anticipated, it has three directions of parallax, and is viewed by intermediate combinations of vertical and horizontal displacement. At this point it should be observed that a generalization can be made regarding viewing n-stereograms, that is, analogous sets of projection pairs, in that hyperobjects of any number of dimensions can be so presented and viewed, with the obvious reduction of perceptual clarity proportionate to the brain's ability to process complex information. The limitation here is not the projective feasibility, but human limitations of such processing. This particular hyperobject may perhaps be better understood as a scanning of a five-dimensional object with a four-dimensional horoptor.

Moving on to the next step, there are the ten coordinate planes of five dimensions (Figure 21). Analogous to the preceding figure of the hypersphere, the great

circles of these planes here define the five-sphere (figure
22).

    '   A little closer look at these surfaces and
hypersurfaces would perhaps help in comprehending them. Let
us consider the projection of a few regular polytopes onto
these surfaces and hypersurfaces. As an example, consider a
spherical triangle as it appears in the stereogram in Figure
23. This is of cource a triangle as it would appear projected onto the surface of a sphere. The next figure shows
a spherical square (Figure 24).

    Let us consider their analogs. There is the projec-
tion of a cube onto the hypersurface of a hypersphere (Figure
25). It should be kept in mind that here, the hyperstereogram presents a curved polytope in a curved three-dimensional space.

    Finally, there is a 5-stereogram of a hypercube
projected onto the four-dimensional curved 4-surface of a
five-sphere (Figure 26). We are thus looking at a curved
figure in four-dimensional curved "space."

    Having considered the surface of constant positive
curvature, let us wind up this study with a brief examination
of a surface of constant negative curvature. Historically,
the pseudosphere is the surface usually considered in this
case. There are, however, some disadvantages to the use of
the pseudosphere. In addition, the non-Euclidean hyperbolic
geometry has been historically defined in relation to a
sphere with an imaginary radius. One of the difficulties
associated with considering such a model has been that the
representation of complex functions requires a four-
dimensional representational matrix. Since we have at our
disposal just such a matrix, it is a relatively simple matter
to adapt it to such a use. Therefore, the following material
will demonstrate a method of representing complex functions
in a perceptually coherent manner, illustrate a method of
representing hyperbolic geometry and complete our general
examination of types of surfaces.

    Because of length limitations, I must assume that
the reader is already familiar with both complex numbers and
the complex plane. Considering the four coordinates of four-
space to consist of two real coordinates and two imaginary
coordinates, I identify them as X, Xi, Y, Yi. These in turn
form six coordinate planes, one real, one imaginary and four
complex. Considering the equation for the circle:
$X^2 + Y^2 = r^2$ and letting $r^2 = -1$, we arrive at the
following equation: $X = \pm i \sqrt{1 + Y^2}$. Determining from this

a range of values of X and Y and expressing them in our
coordinated framework, we arrive at the hyperstereogram of
a surface in complex hyperspace defined by some of the
geodesics of that surface (Figure 27).

Viewed as a four-dimensional figure, that is, a
surface in four dimensions, it is a surface of double
revolution, related to both the Klein bottle and the cylinder
of double revolution. Considering the hyperboloid nature of
its cross-section in the complex planes, it demonstrates an
interesting feature of the complex plane, as each point of
these hyperbolas is the same distance, i.e., the square
root of minus one, from the origin. It follows that
imaginary distances may thus be represented on the complex
plane on other than the coordinate axes. If, in the same
matrix, the radius of the circle is given as the square root
of positive one, then a second hyperbola appears in the
complex plane in relation to the real axis. Finally, the
diagonals of the complex plane, as may be determined by
plugging in appropriate values, are isotropic lines, each
point of which is zero distance from the origin. Needless to
say, this makes it quite clear that Euler's definition of the
absolute value of a complex number: $|z| = + \sqrt{a^2 + b^2}$ when
$z = z + ib$ although useful, and correct as far as it goes,
is incomplete.

To consider another result of this representation,
careful analysis of the descriptive geometry of this
construction reveals that the circles in this diagram
represent lines of exactly the same length, since they occur
in an isocline series and represent parallels of the
hyperbolic plane, if the construction is considered as a
model of that plane. The hyperbolas, which are also straight
lines of the hyperbolic model, all cut a given circle at the
same angle, but cut successive circles at progressively
different angles. This is consistent with classical analysis
of the hyperbolic plane with two extremely important
exceptions. Proofs given for the convergence of parallel
lines in hyperbolic geometry, as will be realized from
examining the literature, are intuitively based on Euclidean
axioms, and are insubstantial. The isocline series of
circles of the representation of our model clearly
demonstrate how lines may be parallel to each other, cut
successively different angles with a "straight" line cutting
through them both, without nevertheless approaching each
other. The two "types" of parallels is quickly resolved when
one realizes that they are related to the two distinct
surfaces that are defined by the imaginary or real values of
the radius. In other words, there are two distinct
hyperbolic planes, one clockwise and the other counter-

clockwise.

Finally, a simple single image of the analog of this surface as a <u>space</u> of constant negative curvature is shown in Figure 28. As you see, it requires six dimensions and is arrived at by using the equation for a sphere with a radius of the square root of negative one.

In conclusion, I believe that there is simply no physical or mathematical conception that is not open to graphic representation, and that visual and verbal thought are isomorphic to each other. Although there is much more material which could have been considered here, I hope this chapter has inspired you to share this belief.

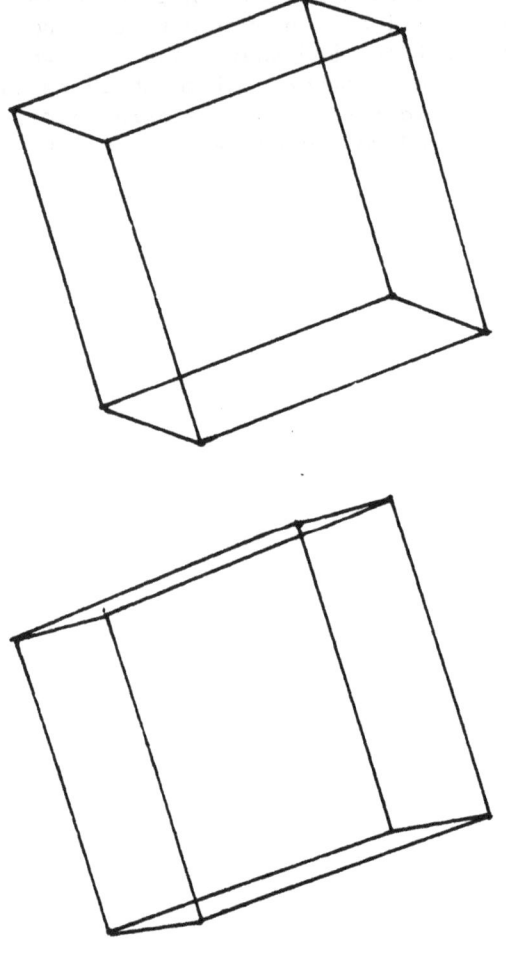

FIG. 1

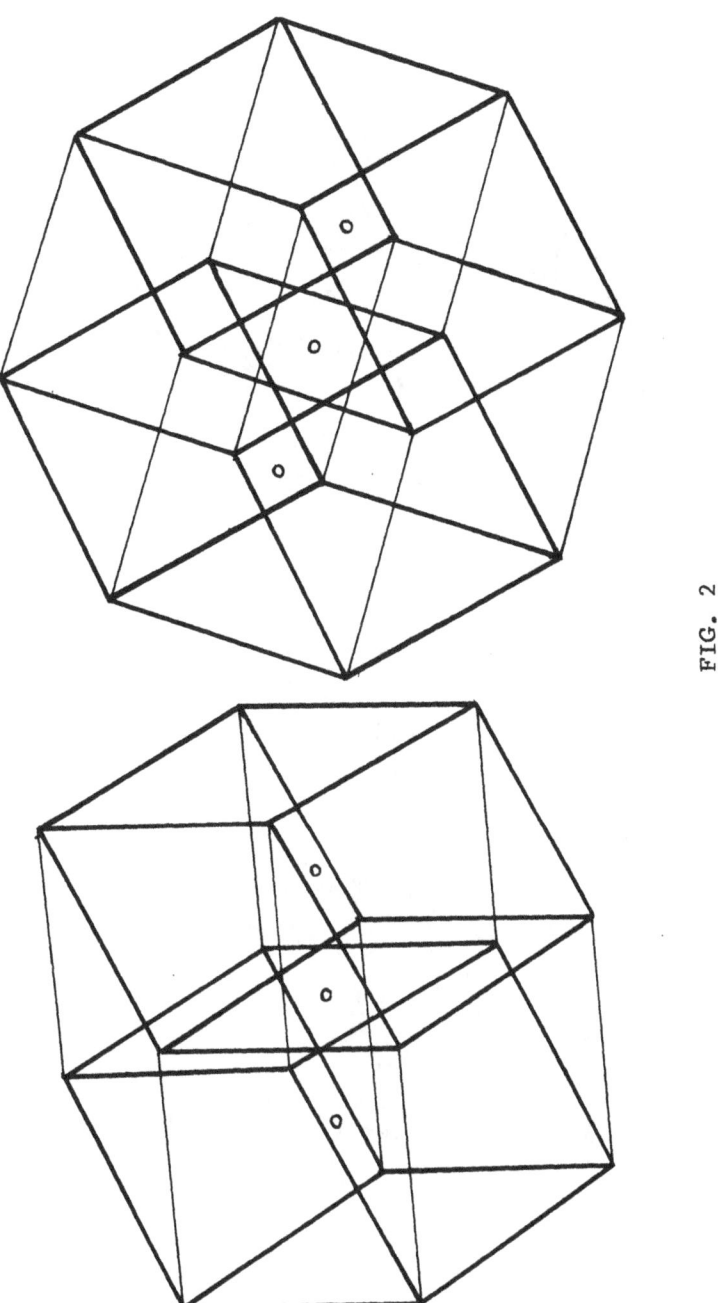

FIG. 2

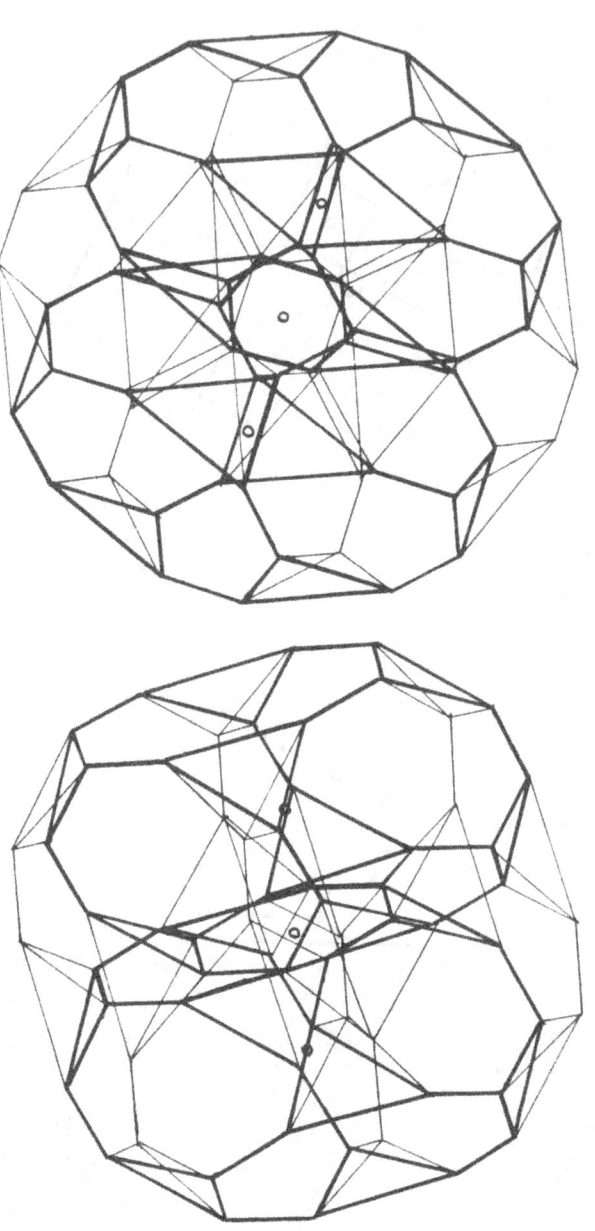

FIG. 3

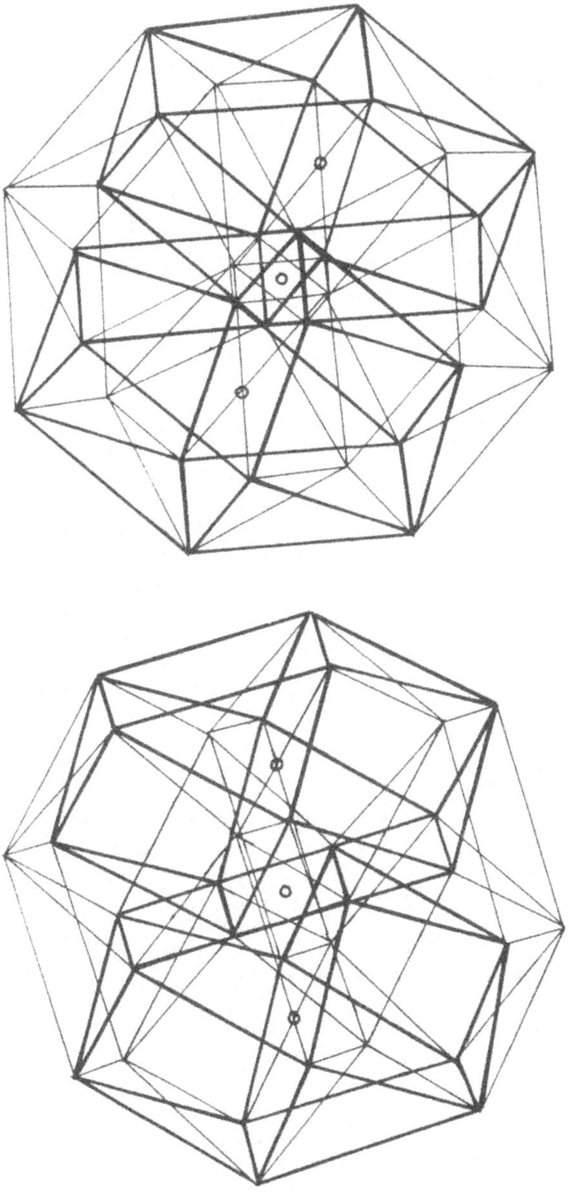

FIG. 4

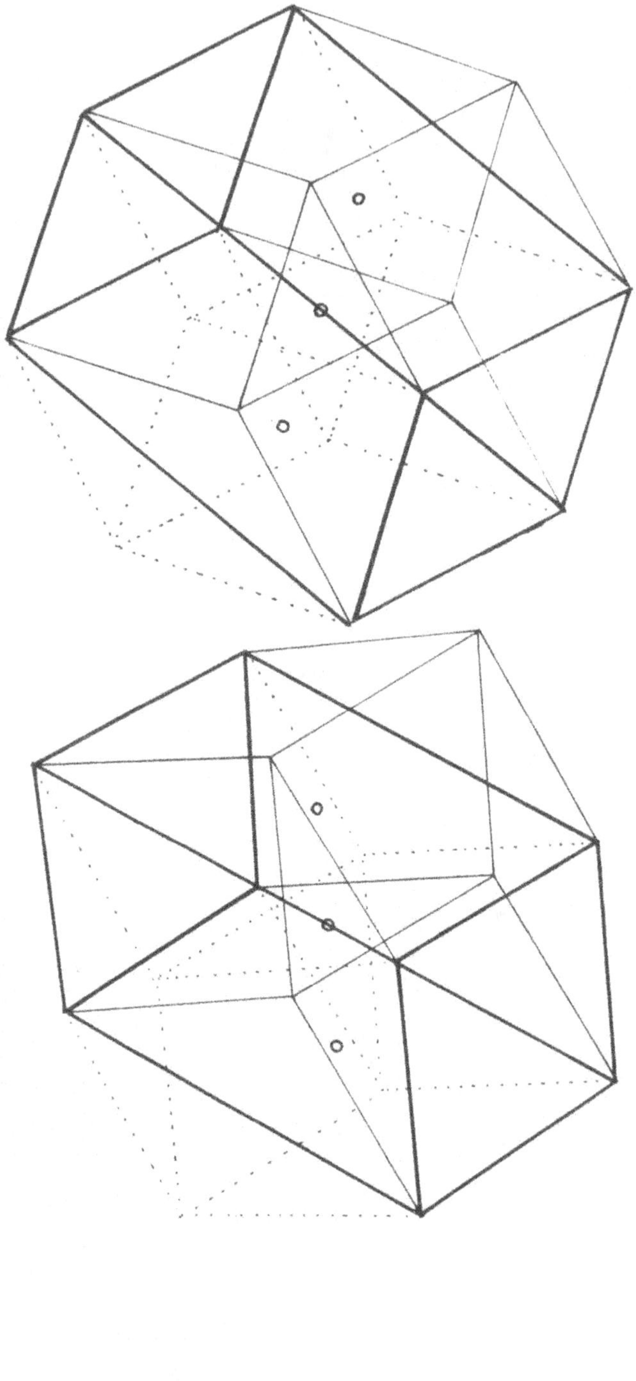

FIG. 5

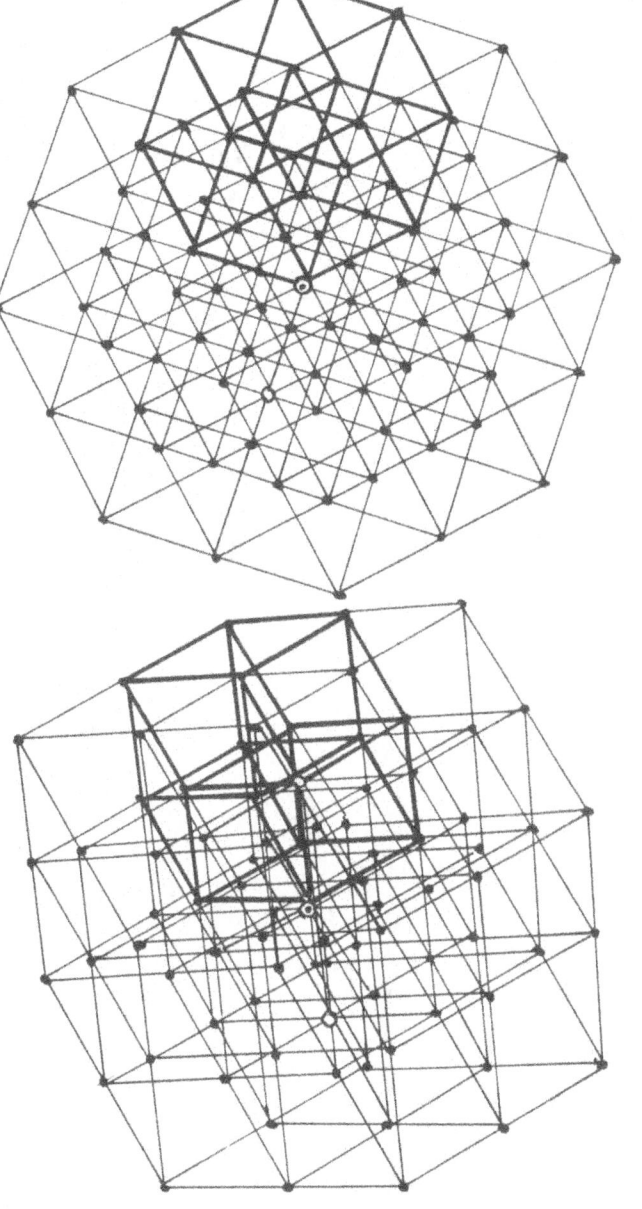

FIG. 6

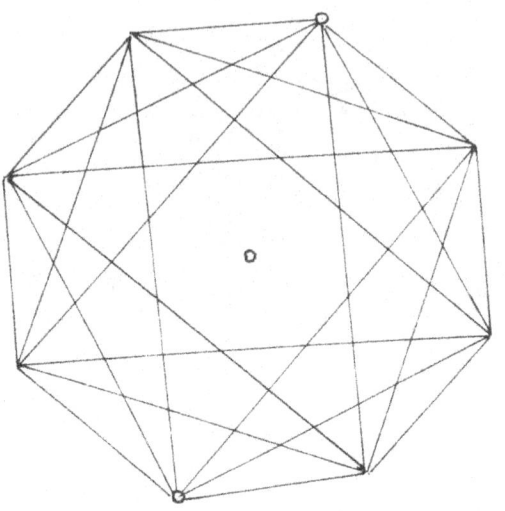

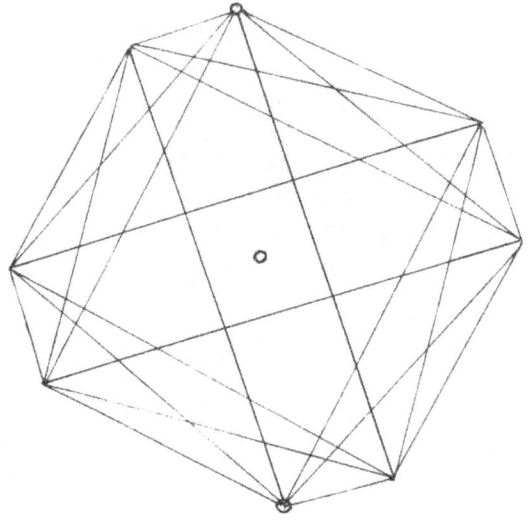

FIG. 7

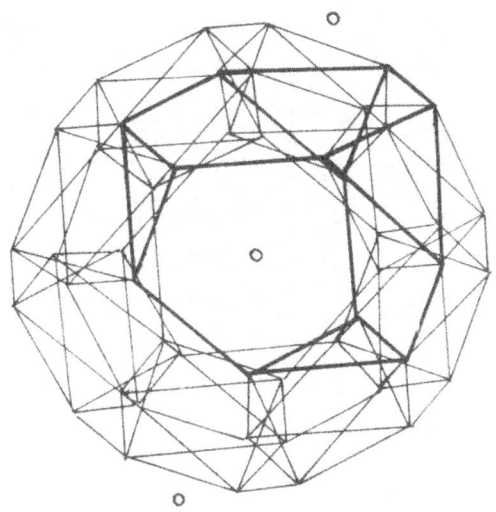

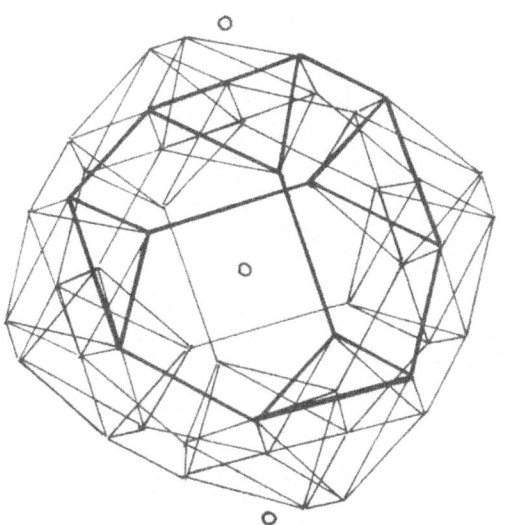

FIG. 8

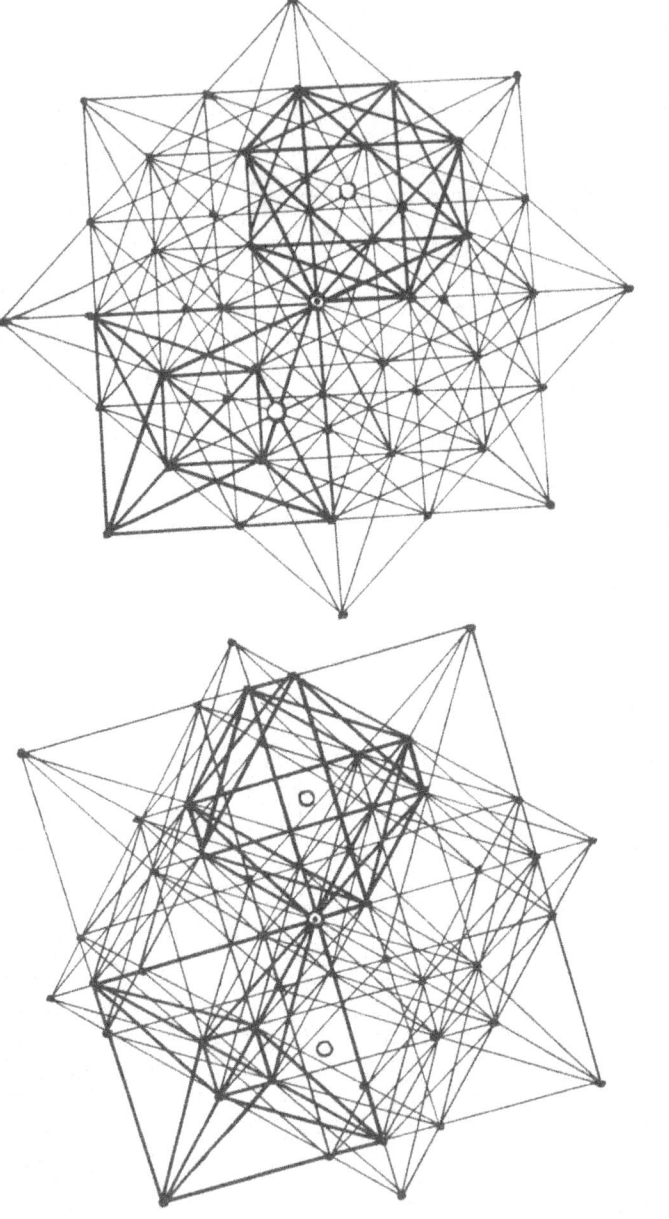

FIG. 9

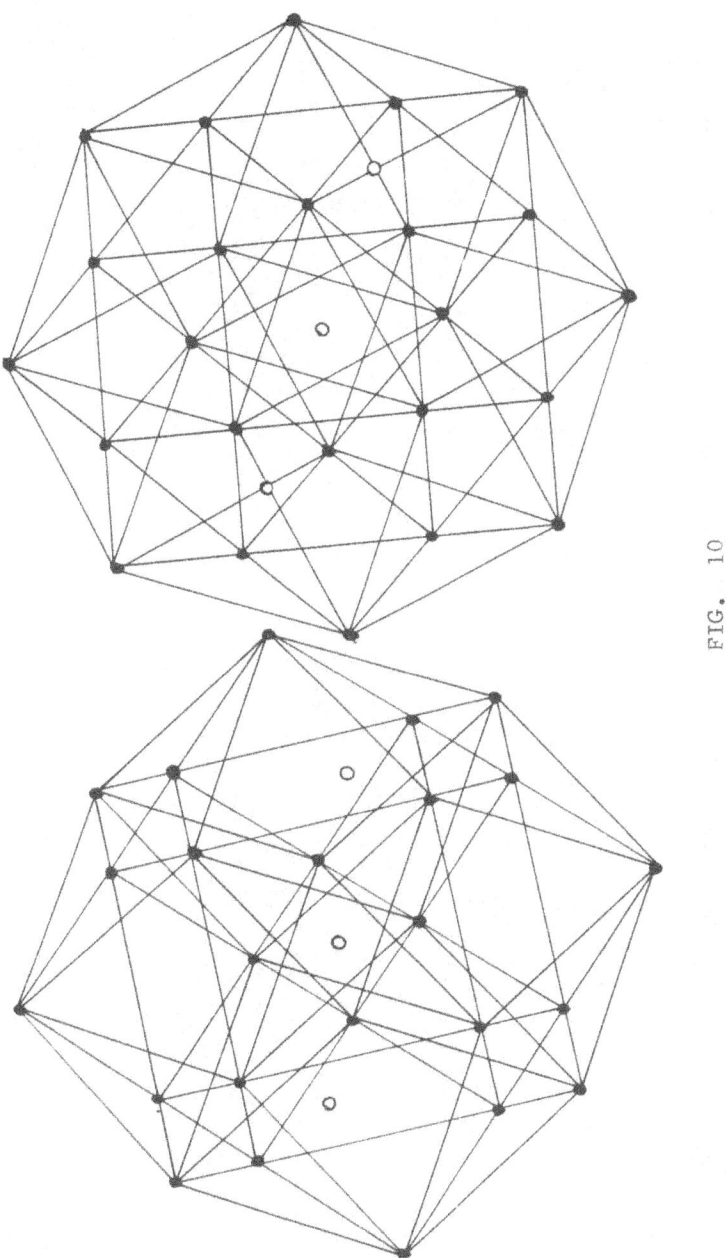

FIG. 10

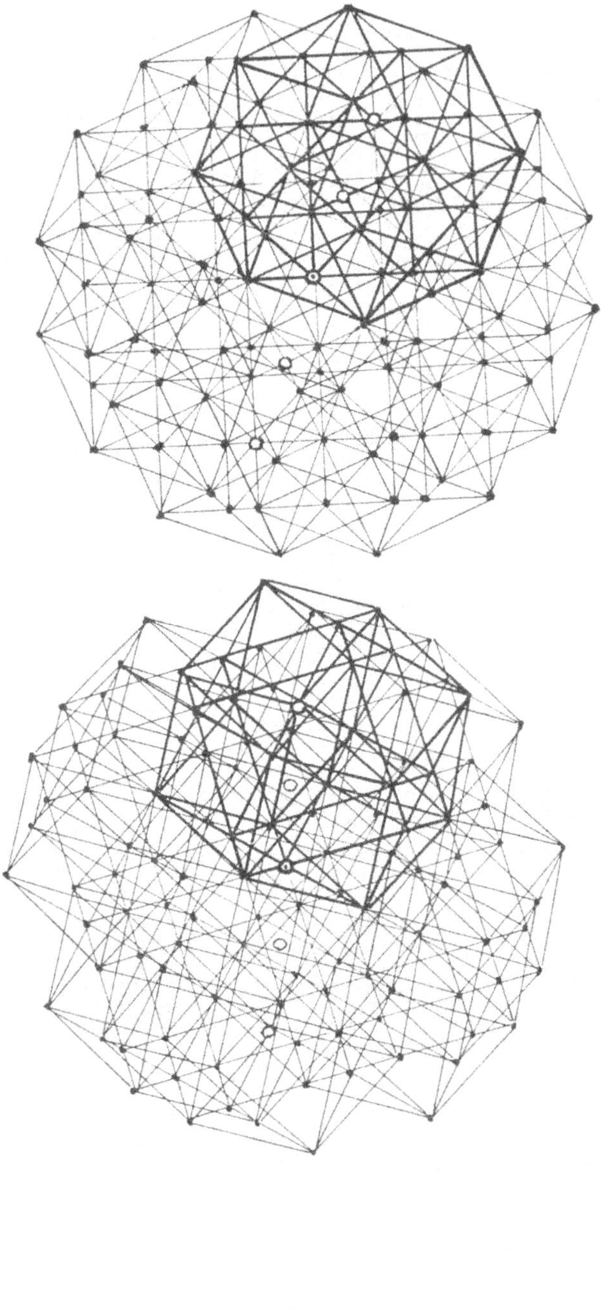

FIG. 11

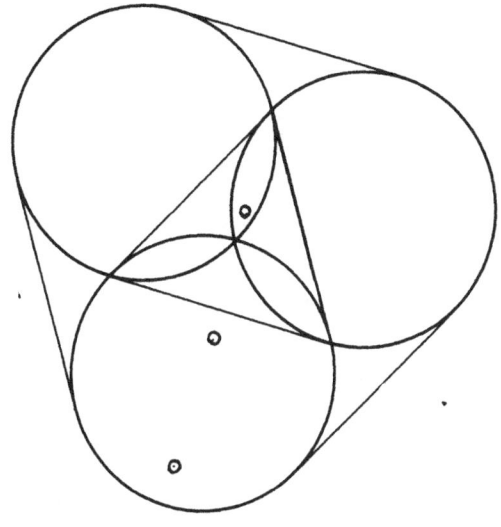

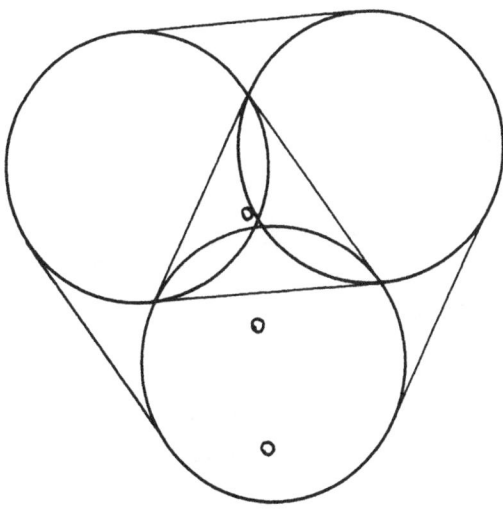

FIG. 12

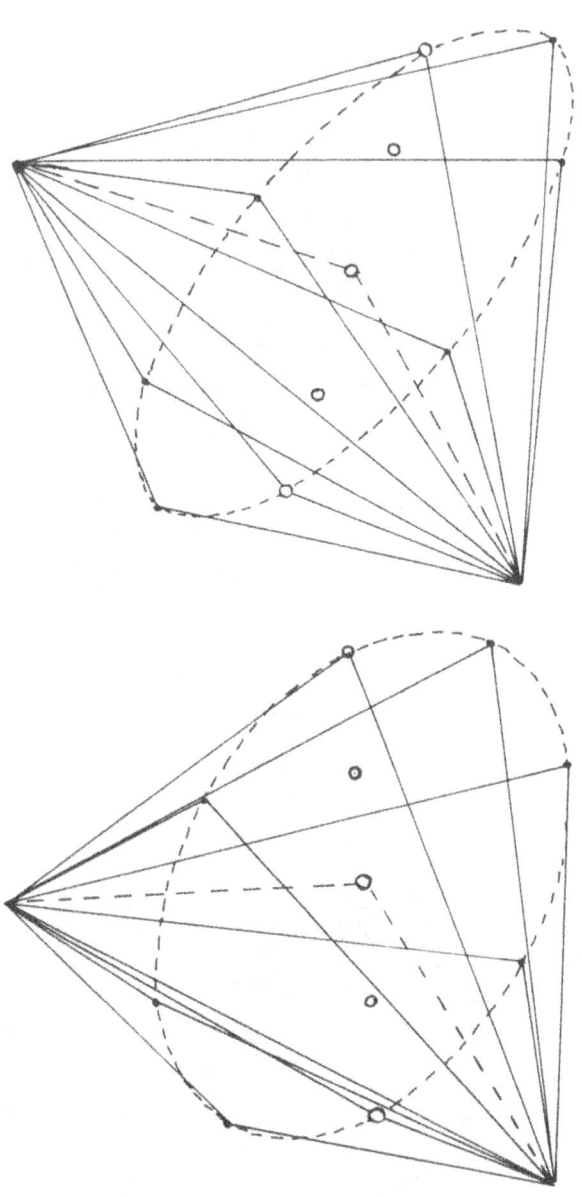

FIG. 13

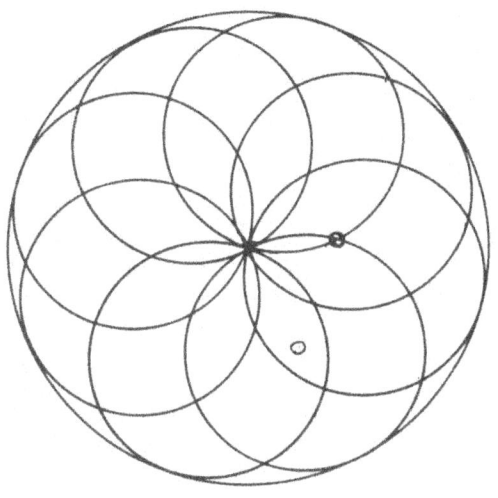

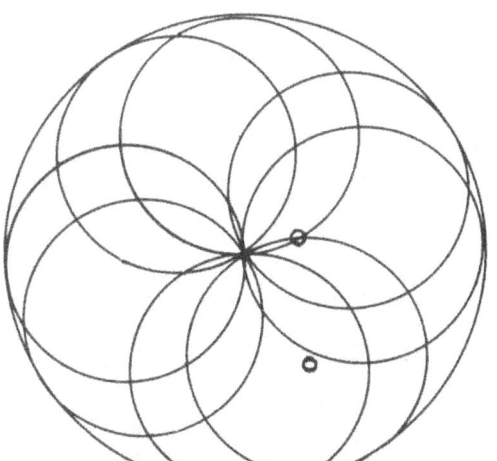

FIG. 14

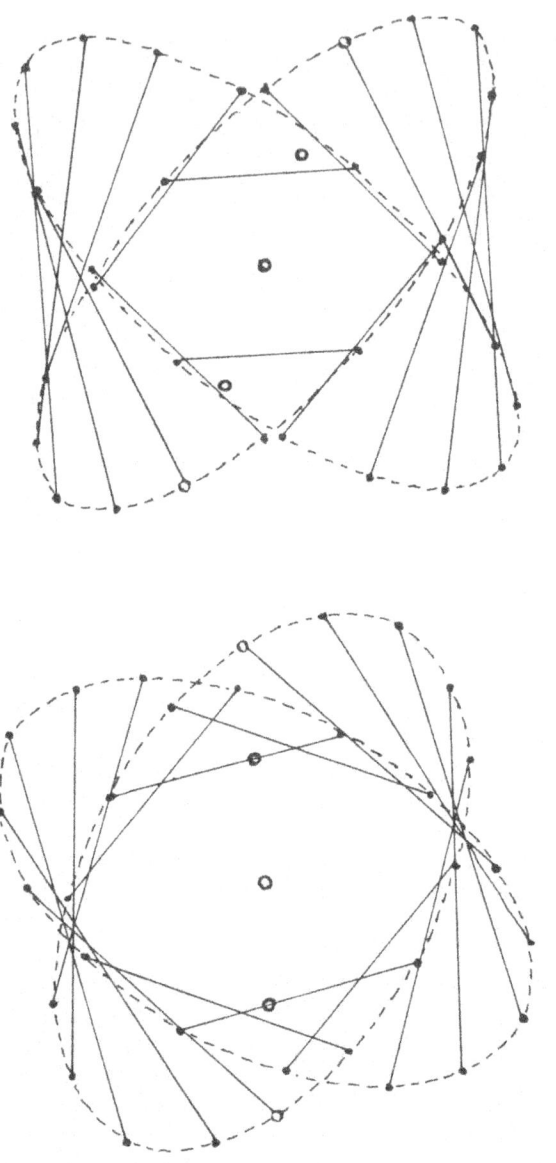

FIG. 15

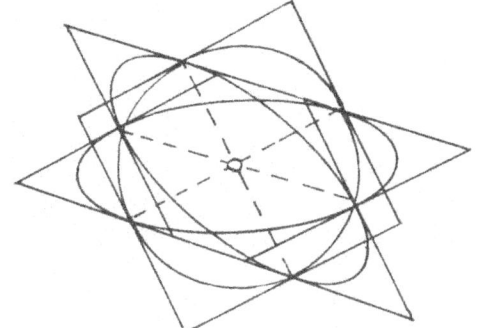

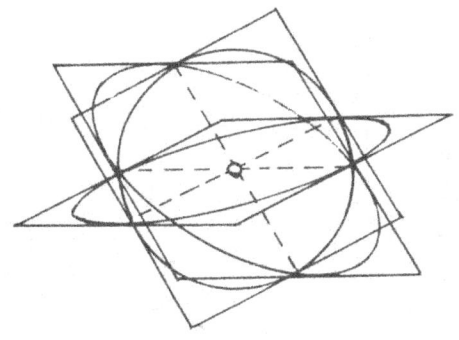

FIG. 16

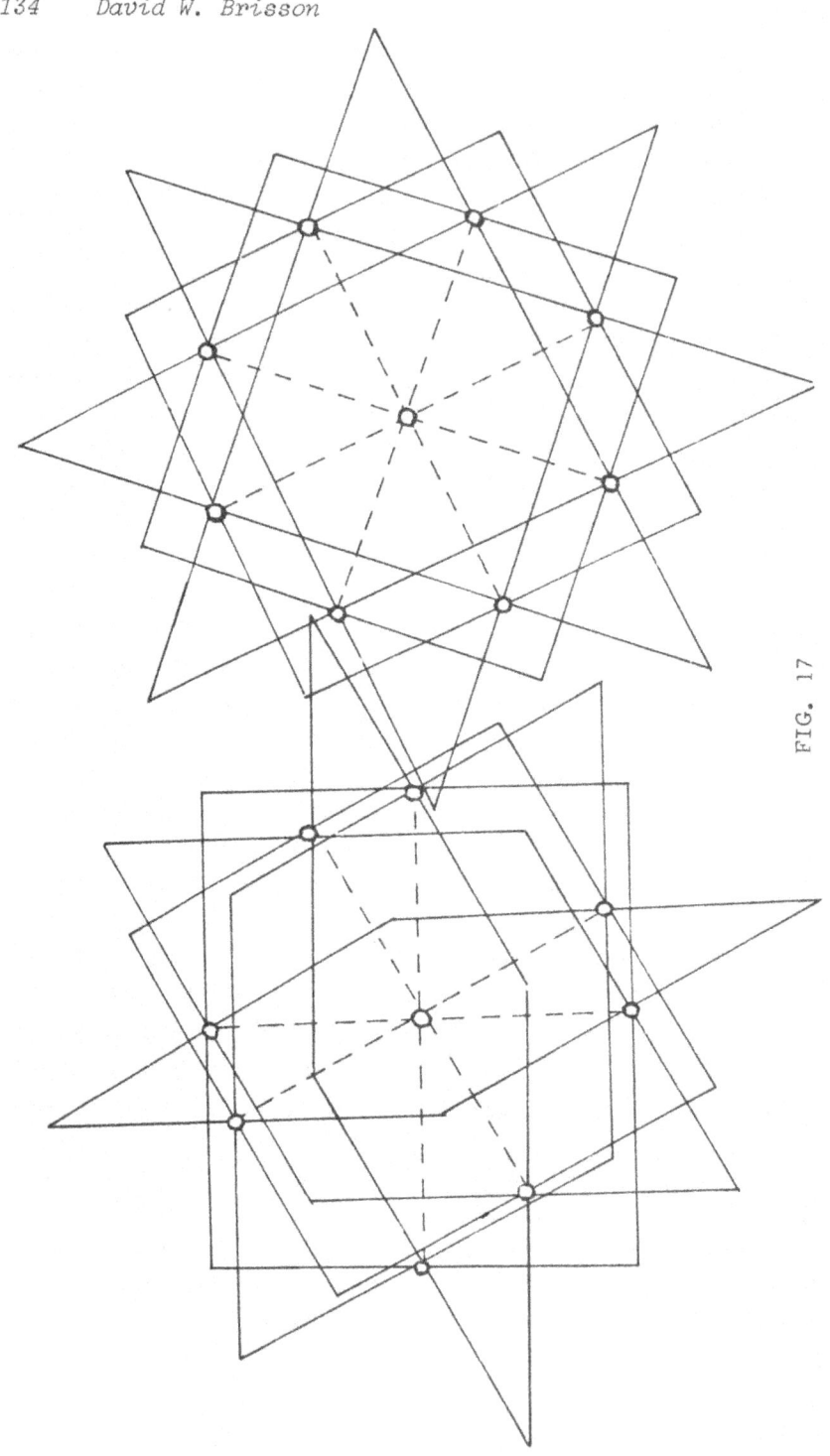

FIG. 17

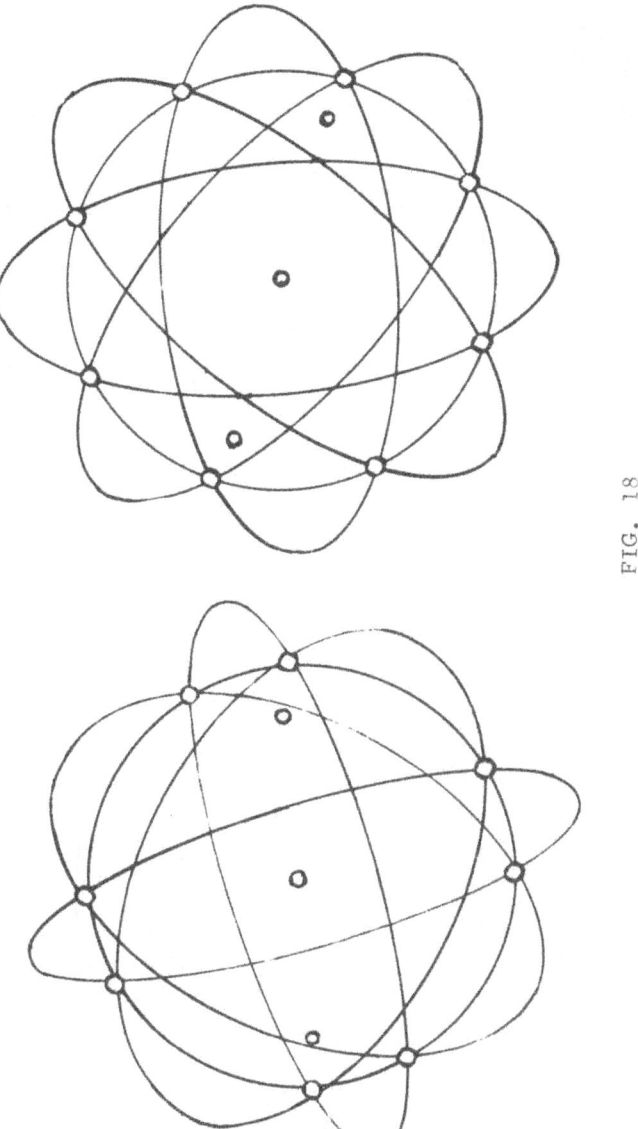

FIG. 18

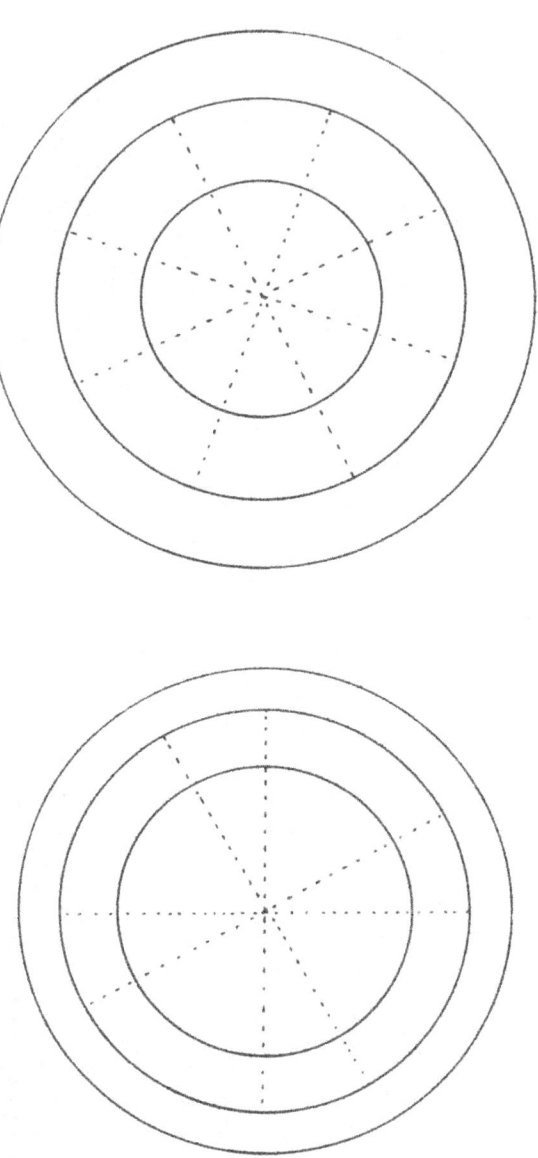

FIG. 19

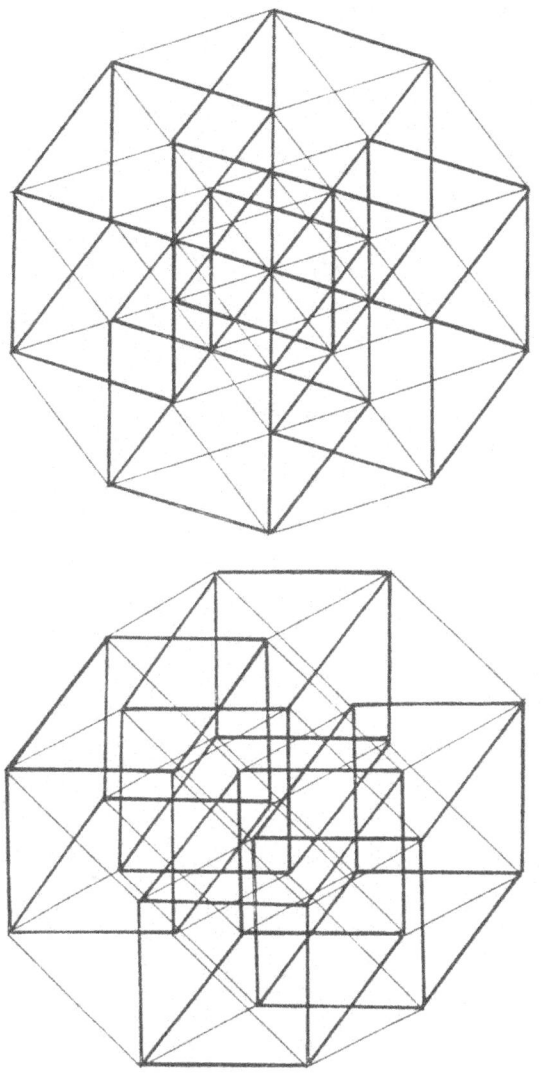

FIG. 20

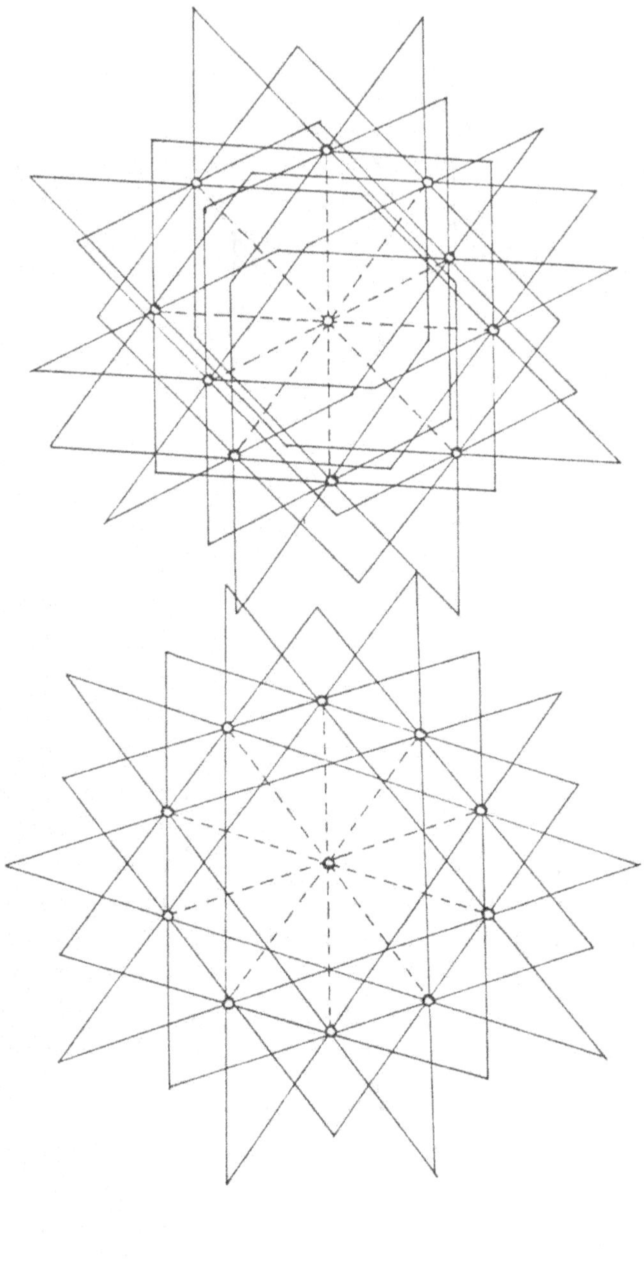

FIG. 21

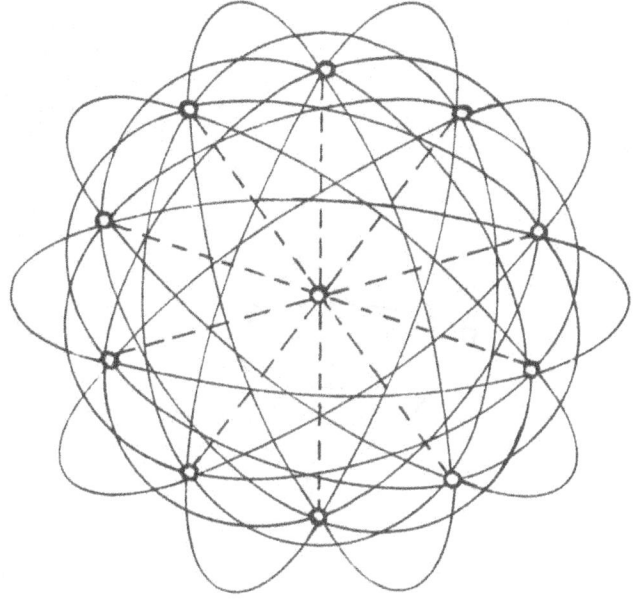

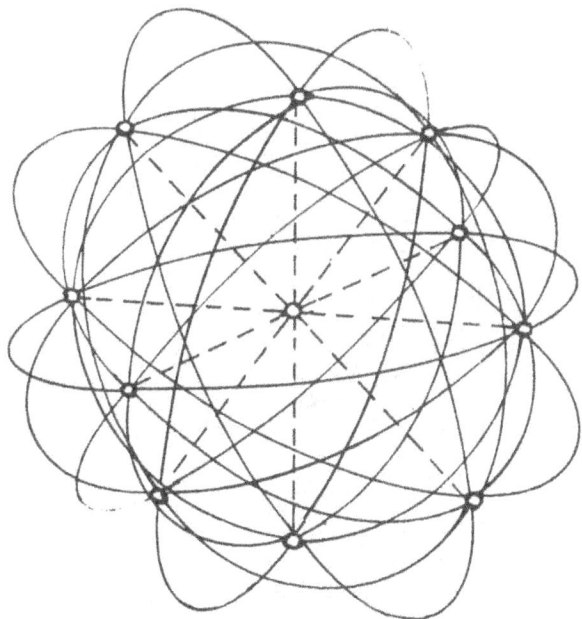

FIG. 22

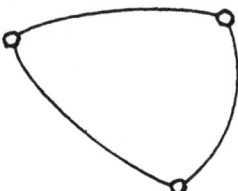

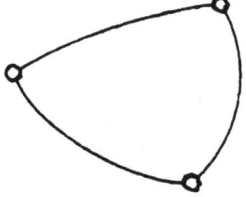

FIG. 23

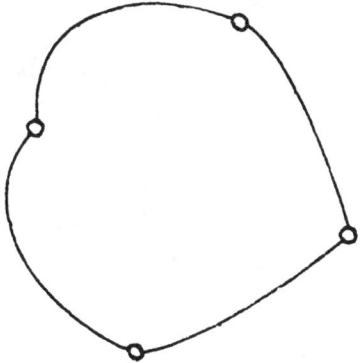

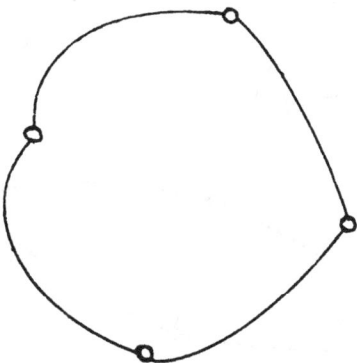

FIG. 24

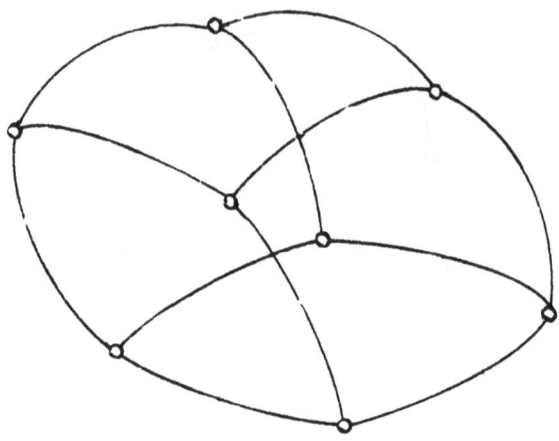

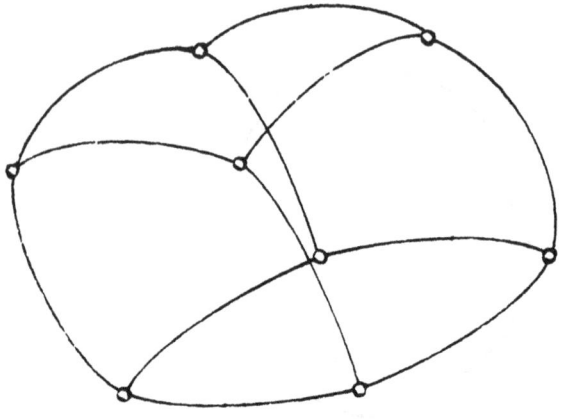

FIG. 25

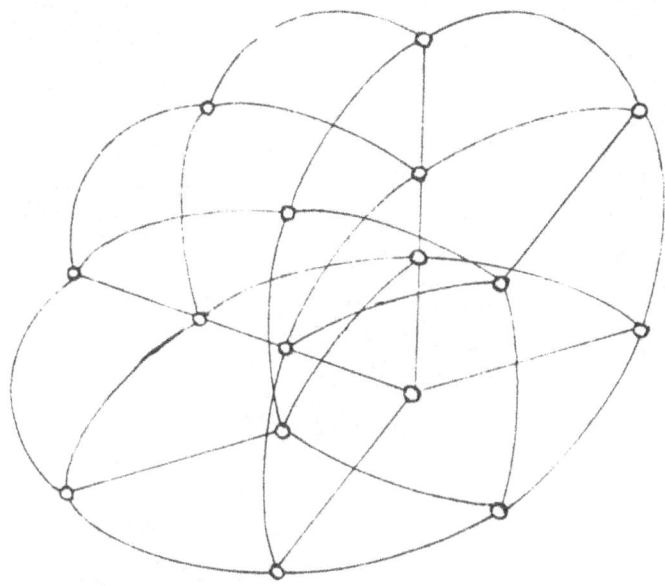

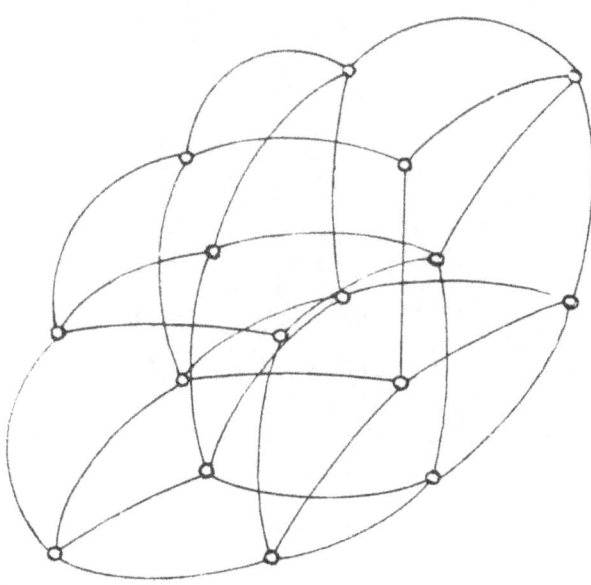

FIG. 26

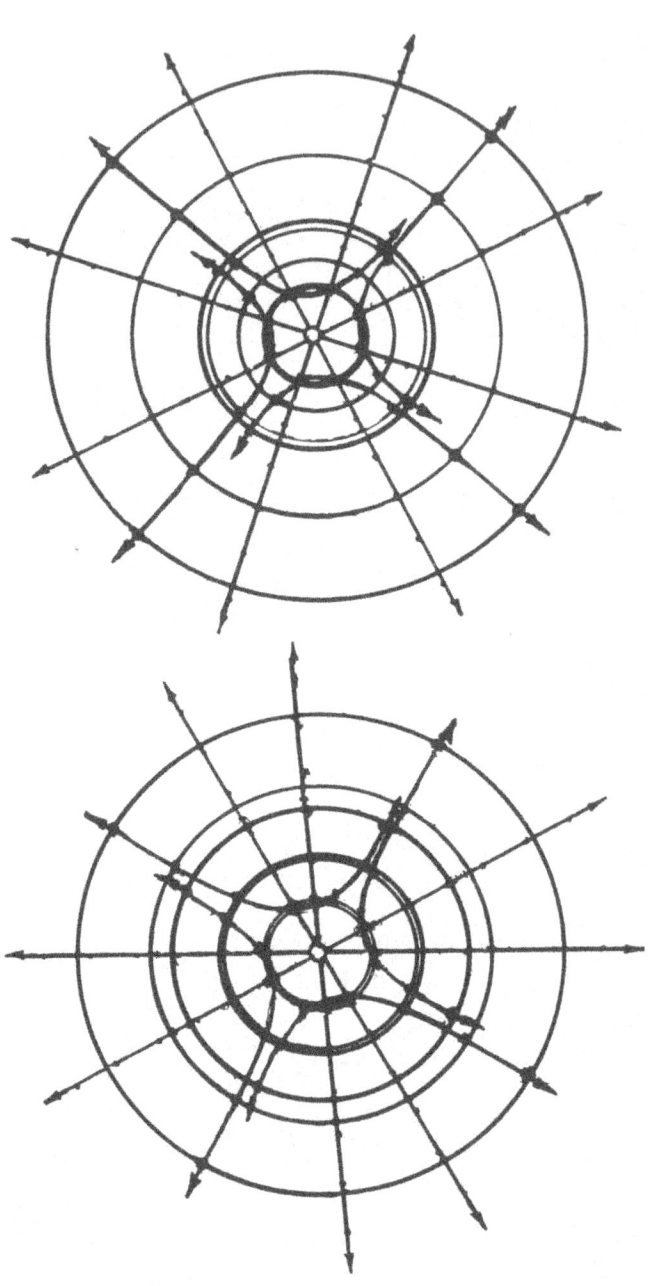

FIG. 27

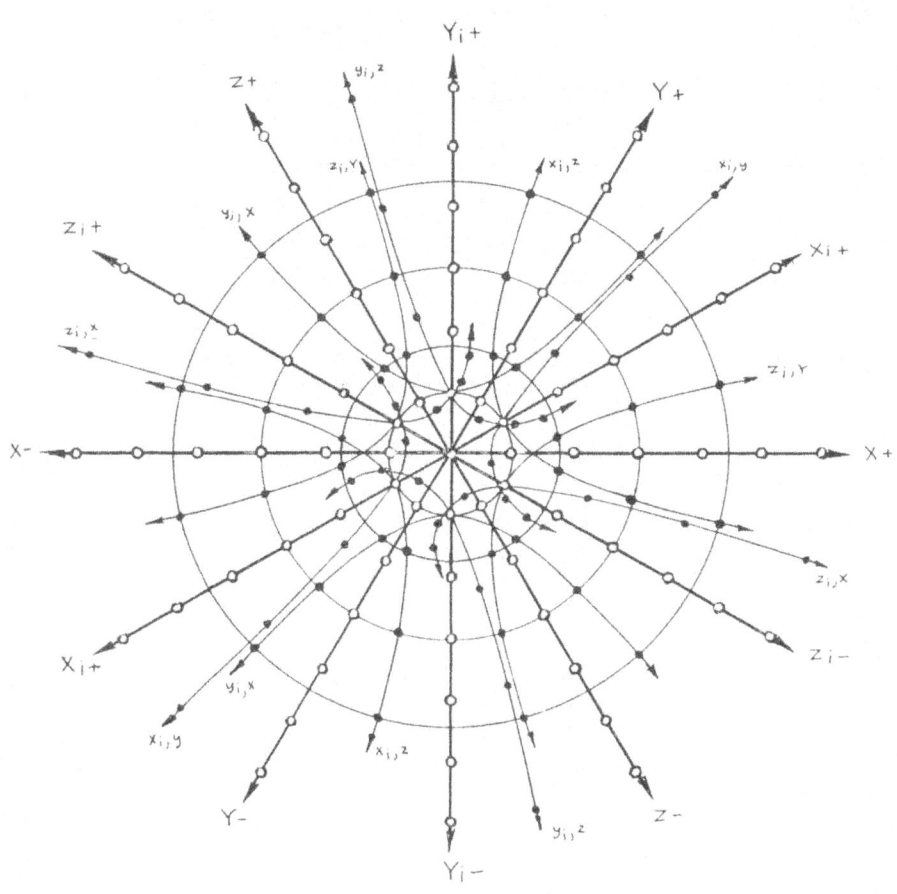

FIG. 28

Displaying n-Dimensional Hyperobjects by Computer

A. Michael Noll

A digital computer and automatic plotter have been
used to generate three-dimensional stereoscopic movies of the
three-dimensional parallel and perspective projections of
four-dimensional hyperobjects rotating in four-dimensional
space. The observed projections and their motions were a
direct extension of three-dimensional experience, but no
profound "feeling" or insight into the fourth spatial
dimension was obtained. The technique can be generalized
to n-dimensions and applied to any n-dimensional hyperobject or hypersurface.

Introduction

In his now classic book on Flatland, Edwin Abbott (1)
describes the social order resulting in a world restricted to
two spatial dimensions. The inhabitants of this world are
completely unable to visualize a third spatial dimension and
are therefore thoroughly baffled by the weird distortions of
the two-dimensional projections into their world of simple
three-dimensional objects. Man finds himself in a similar
state of puzzlement concerning spatial dimensions higher than
three, and really never even knows whether he might be
witnessing the three-dimensional projection of some higher-
dimensional event. When man does not comprehend he sometimes
gives religious significance, and therefore not surprisingly
a fifth dimension has even been proposed as "the ultimate
spiritual essence" (8).

The mathematics and projective geometry of three-
dimensional space can be generalized to any number of

*First published in Communications of the ACM 10, 469 (1967)
as "A Computer Technique for Displaying n-Dimensional Hyperobjects." Copyright 1967, Association for Computing Machinery,
Inc. Reprinted by permission.

dimensions so that an n-dimensional hyperobject can be
mathematically projected into an (n- 1)-dimensional space.
Such projection could be applied repetitively until finally a
three-dimensional object representing the successive
projections of an n-dimensional hyperobject is obtained. If
desired, the hyperobject might move in n-dimensional space so
that its three-dimensional projection would not be
stationary. A relatively simple form of motion is rotation
of the hyperobject in n-dimensional space.

This paper is a review of the mathematics for two
types of projection of n-dimensional hyperobjects and for n-
dimensional rotation. Any n-dimensional hyperobject could
then be manipulated mathematically by a digital computer.
The final three-dimensional projection of the rotating
hyperobject could be drawn automatically on a computer-
controlled visual display device as a stereoscopic movie.

As an example of this technique, a computer technique
for generating three-dimensional movies of the perspective
projections into three-dimensional space of four-dimensional
hyperobjects is described. Now, like the inhabitants of
Flatland, we too are puzzled by the strange distortions of
the projection into three-dimensional space of a rotating but
rigid four-dimensional hyperobject. Using intuition to
extend to four dimensions our knowledge of the type of
distortions resulting from three-dimensional perspective
projection, it is possible to explain the distortions but
still impossible to visualize the rigid four-dimensional
hyperobject.

Rotation

Throughout this paper, bold-face lower-case letters
will represent column vectors while matrices will be
represented by bold-face capital letters. The column vector
$x = (x_1, x_2, \cdots, x_n)^t$ represents any point in n-dimensional
space where the superscript t indicates transposition.

In n-dimensional space the simplest rotation is in a
two-dimensional plane. If rotation is in the plane of x_a
and x_b (the a-b plane), then the rotation matrix $R_{ab}(\alpha)$
has the elements

$$r_{ii} = 1 \quad \text{except } r_{aa} = r_{bb} = \cos \alpha, \quad (1)$$
$$r_{ij} = 0 \quad \text{except } r_{ab} = -r_{ba} = -\sin \alpha.$$

For example, the rotation matrix for a rotation through an
angle α in the 2-4 plane in five-dimensional space is

$$
R_{24}(\alpha) \;=\; \begin{pmatrix} 1 & 0 & 0 & 0 & 0 \\ 0 & \cos\alpha & 0 & -\sin\alpha & 0 \\ 0 & 0 & 1 & 0 & 0 \\ 0 & \sin\alpha & 0 & \cos\alpha & 0 \\ 0 & 0 & 0 & 0 & 1 \end{pmatrix} \qquad (2)
$$

The n-dimensional rotation specified by eq. (1) is called a two-dimensional plane rotation since only those co-ordinates of a vector in the two-dimensional plane determined by the axes x_a and x_b are changed. Thus, in three-dimensional space, rotation about the x_3 axis would be called rotation in the x_1-x_2 plane. In spaces of higher than three dimensions, rotation about an axis is meaningless in terms of eq. (1) since a multitude of nonparallel two-dimensional planes are all perpendicular to the same axis. For example, in four-dimensional space the x_1-x_2, x_1-x_3, and x_2-x_3 planes are all perpendicular to the x_4-axis.

Any n-dimensional rotation matrix can be written as the product of $n(n-1)/2$ two-dimensional-plane n-dimensional rotation matrices (5). Thus, e.g., in four-dimensional space

$$
R \;=\; R_{12}(\alpha_1)R_{13}(\alpha_2)R_{14}(\alpha_3)R_{23}(\alpha_4)R_{24}(\alpha_5)R_{34}(\alpha_6). \quad (3)
$$

Projection

The most common form of projection in three-dimensional space is the perspective projection of an object as seen by each of our two eyes. A perspective projection is produced graphically, as shown in Figure 1, by first choosing a viewing point from which to view the object. A two-dimensional plane is then placed between the object and the the viewing point. Straight lines are drawn from the object to the viewing point; their intersections with the plane are the perspective projection of the object. This procedure can be extended as follows to n-dimensional objects.

As shown in Figure 2, the point $P = (X_1, X_2, \cdots, X_n)^t$ in the n-dimensional space with axes x_1, x_2, $\cdots$, x_{n-1}, x_n is to be perspectively projected onto the $(n-1)$-dimensional hyperplane defined by $x_n = F$. For simplicity, the viewing point v is located a distance R along the x_n-axis. A straight line is drawn from p to v, and its intersection p' with the $(n-1)$-dimensional hyperplane is the desired

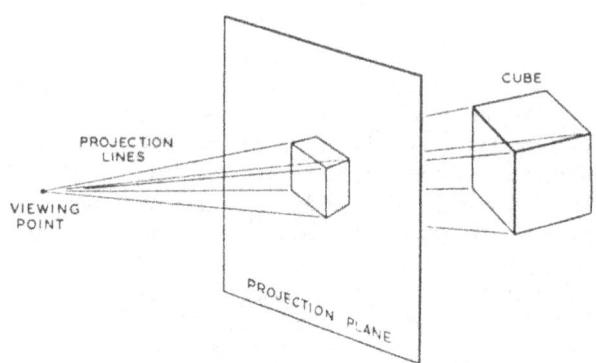

FIG. 1. Perspective projection of a three-dimensional cube onto a two-dimensional plane

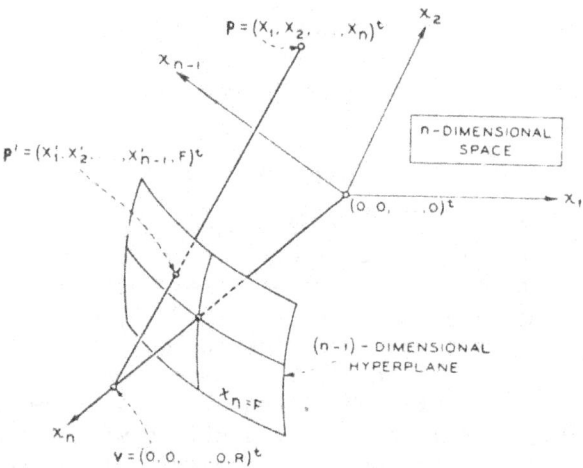

FIG. 2. Perspective projection of a point in n-dimensional space onto an $(n-1)$-dimensional hyperplane

perspective projection:

$$\mathbf{p}' = \begin{pmatrix} X_1' \\ X_2' \\ \vdots \\ X_{n-1}' \\ X_n' \end{pmatrix} = \begin{pmatrix} \dfrac{(R-F)X_1}{R-X_n} \\ \dfrac{(R-F)X_2}{R-X_n} \\ \vdots \\ \dfrac{(R-F)X_{n-1}}{R-X_n} \\ F \end{pmatrix} \qquad (4)$$

The nth coordinate of $\mathbf{p}'$ is constant since $\mathbf{p}'$ lies in the $(n-1)$-dimensional hyperplane. Accordingly, $\mathbf{p}$ can also be represented as a $(n-1)$-dimensional vector mathematically derived as the perspective projection of its counterpart in the n-dimensional space:

$$\mathbf{p}'_{n-1} = \begin{pmatrix} \dfrac{(R-F)X_1}{R-X_n} \\ \dfrac{(R-F)X_2}{R-X_n} \\ \vdots \\ \dfrac{(R-F)X_{n-1}}{R-X_n} \end{pmatrix} \qquad (5)$$

Another type of projection is derived from perspective projection by choosing the viewing point at infinity, i.e., $v = (0, 0, \cdots, \infty)^t$. Since the projection lines are all parallel in the limit, this is commonly called parallel projection. By taking the limit of eq. (4) as $R \to \infty$, the parallel projection $\mathbf{q}'$ of the n-dimensional point $p = (X_1, X_2, \cdots, X_n)^t$ is

$$\mathbf{q}' = \begin{pmatrix} X_1 \\ X_2 \\ \vdots \\ X_{n-1} \\ F \end{pmatrix} \qquad (6)$$

Thus, the parallel projection is identical with the original point in the first $(n-1)$ dimensions.

Hyperobjects

A hyperobject in n-dimensional space can be represented as straight line segments connecting an ordered set of points (m in number). Similarly, an n-dimensional hypersurface can be depicted visually as a finite set of points randomly scattered over its surface. In either case, the n-dimensional hyperobject or hypersurface can be specified as a set of n-dimensional vectors which, if desired, might be combined together as the columns of a matrix. Thus, an n-dimensional hyperobject or hypersurface can be represented as an $n \times m$ matrix H given by

$$H = \begin{pmatrix} Y_1(1) & Y_1(2) & \cdots & Y_1(m) \\ Y_2(1) & Y_2(2) & \cdots & Y_2(m) \\ \vdots & \vdots & & \vdots \\ Y_n(1) & Y_n(2) & \cdots & Y_n(m) \end{pmatrix}. \tag{7}$$

The mathematical restriction on the Y's or the algorithm used in calculating them determines the hyperobject or hypersurface represented by the matrix H. For example, if

$$\sum_{i=1}^{n} (Y_i(j) - C_i)^2 = \rho^2 \tag{8}$$

for all $j = 1, \ldots, m$, then the points all lie on the surface of an n-dimensional hypersphere with center at $(C_1, C_2, \ldots, C_n)^t$.

If the line representation of a hyperobject with disjoint portions is desired, then the disjoint portions of the hyperobject must be specified so as not to be connected together with straight lines. Also, even if the hyperobject is not disjoint, certain line segments might have to be treated disjointly if the restriction is imposed that no line be drawn twice. For example, the 12 edges of a three-dimensional cube can not be drawn as a connected line without drawing some edges more than once.

Computer Technique

Since our habitation is restricted to a maximum of three spatial dimensions, we are unable to visualize a fourth much less a higher spatial dimension. We are able to perceive three-dimensional depth as a result of the slightly different images seen by our eyes. The illusion of depth can be created by viewing stereoscopically a pair of perspective two-dimensional pictures, but such perspectives are very tedious to calculate and draw. However, computer techniques are presently available for calculating and automatically plotting the left eye and right-eye images of some three-dimensional object (6). If desired, a three-dimensional movie can be generated in this manner using the computer and automatic plotter to generate a sequence of pictures. Thus, if an n-spatial-dimensional object is mathematically projected onto three dimensions, the computer can produce the

required drawings to obtain a three-dimensional depth effect.
A movie can be produced by simply choosing to rotate the
hyperobject in n-dimensional space. Although this procedure
is generally applicable to n-dimensions, the details that
follow will describe the actual implementation to four-
dimensional hyperobjects and hypersurfaces.

The computer program performs the following
functions. First, the object matrix H specifying the
desired hyperobject is read into the computer from punched
cards. The hyperobject is then rotated in four-dimensional
space to a new orientation with object matrix

$$Y = R_{ab}(\alpha)H \qquad (9)$$

for a single plane rotation or by

$$Y = R_{ab}(\alpha)R_{cd}(\beta)R_{ef}(\gamma)H \qquad (10)$$

for a succession of three plane rotations. The rotated
hyperobject Y is projected into three-dimensional space
either by perspective projection given by

$$
\begin{pmatrix}
X_1(1) & \cdots & X_1(m) \\
X_2(1) & \cdots & X_2(m) \\
X_3(1) & \cdots & X_3(m)
\end{pmatrix}
$$

$$
= (R - F)
\begin{pmatrix}
\dfrac{Y_1(1)}{R - Y_4(1)} & \cdots & \dfrac{Y_1(m)}{R - Y_4(m)} \\[2ex]
\dfrac{Y_2(1)}{R - Y_4(1)} & \cdots & \dfrac{Y_2(m)}{R - Y_4(m)} \\[2ex]
\dfrac{Y_3(1)}{R - Y_4(1)} & \cdots & \dfrac{Y_3(m)}{R - Y_4(m)}
\end{pmatrix} \qquad (11)
$$

or by parallel projection given by

$$
\begin{pmatrix}
X_1(1) & \cdots & X_1(m) \\
X_2(1) & \cdots & X_2(m) \\
X_3(1) & \cdots & X_3(m)
\end{pmatrix}
=
\begin{pmatrix}
Y_1(1) & \cdots & Y_1(m) \\
Y_2(1) & \cdots & Y_2(m) \\
Y_3(1) & \cdots & Y_3(m)
\end{pmatrix} \qquad (12)
$$

The stereoscopic pair of two-dimensional perspective
projections of the three-dimensional projection of the
hyperobject are calculated by the computer and automatically
plotted on a single frame of film. The hyperobject is then
rotated through an incremental angle, and the various
projections from four dimensions to three dimensions and from
three dimensions to a pair of two-dimensional projections are
repeated finally resulting in yet another frame of the movie.

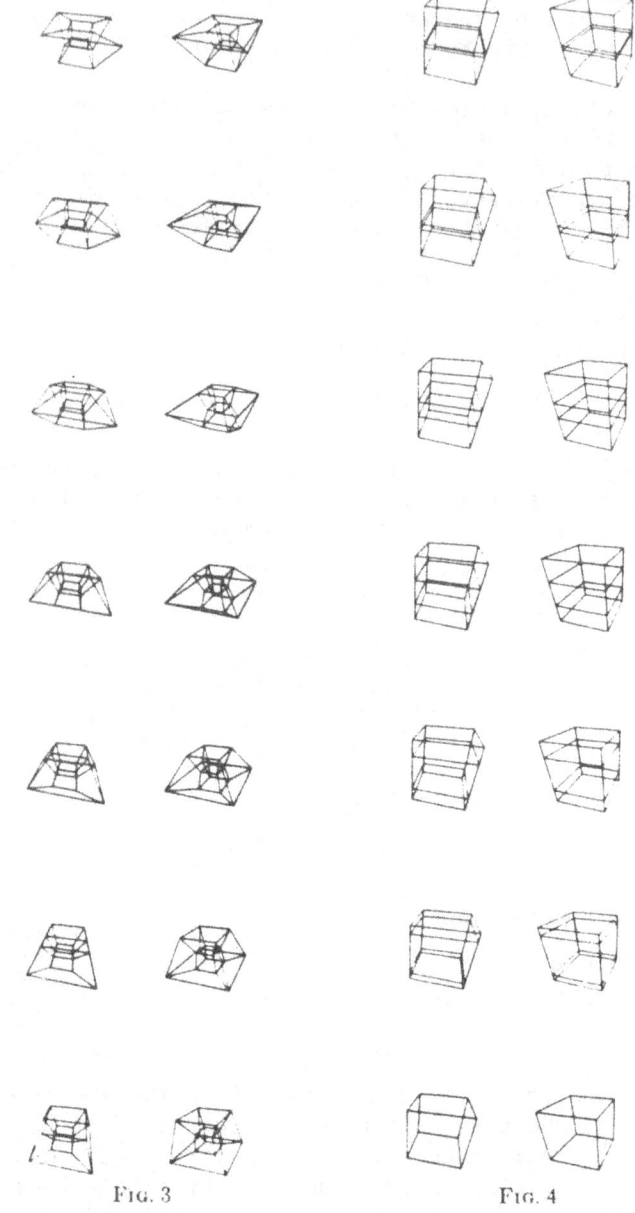

Fig. 3   Fig. 4

Fig. 3. Selected frames from a computer-generated three-dimensional movie showing the three-dimensional perspective projection

Examples

The hypercube is the n-dimensional generalization of
either a two-dimensional square or a three-dimensional cube.
It is bounded by pairs of parallel (n - 1)-dimensional
hyperplanes which are all the same distance apart. The
three-dimensional cube is bounded by three pairs of two-
dimensional faces or squares while the four-dimensional
hypercube is bounded by four pairs of three-dimensional
hyperfaces which now are cubes. The n-dimensional hypercube
has 2^n vertices and $n \cdot 2^{n-1}$ edges so that a four-
dimensional hypercube has 16 vertices and 32 edges.

The 16 vertices specifying a four-dimensional
hypercube were calculated and ordered into 33 points which
when connected sequentially by straight lines would produce
the hypercube's 32 edges. These points formed the object-
matrix specification of the hypercube and the computer then
produced three-dimensional movies of the parallel and
perspective projections of the hypercube rotating in four-
dimensional space. Selected frames from the movie of the
perspective projection are shown in Figure 3.

The perspective projection of a four-dimensional
hypercube is a cube within a cube with corresponding vertices
connected together. This and the motion caused by rotation
can better be understood by analogy with a three-
dimensional cube. The perspective projection of a cube is a
square within a square with pairs of edges connected
together, because the face, here a square, closest to the
viewing point will appear largest. Similarly, the
hypercube's hyperface (now a cube) closest to the viewing
point will appear largest so that a cube within a cube is
obtained.

If a cube rotates in a two-dimensional plane
perpendicular to a line passing through both the origin and
the viewing point, the perspective projection simply rotates
as a whole. If, however, this line does not pass through the
viewing point, then the position of the faces changes so that
the projections of the faces change their size as the cube
rotates. The four-dimensional extension of this is that the
three-dimensional cube within a cube similarly rotates as a
whole if the line passing through both the origin and the
viewing point is perpendicular to the rotation plane.
However, when this line is not perpendicular to the rotation
plane, the cubes change their size as they rotate so that the
hyperface (cube) closest to the viewing point is always
largest. In the actual movie, the viewing point is situated

on the x_4-axis, and three different matrix transformations are used for the rotations: (1) three complete revolutions in the 1-3 plane, i.e., $R_{13}(\alpha)$; (2) three complete revolutions in the 2-4 plane, i.e., $R_{24}(\alpha)$, and (3) three successive matrix transformations, i.e., $R_{23}(\alpha)R_{13}(\beta)R_{31}(\gamma)$.

For the parallel projection from four dimensions to three dimensions there are no perspective distortions, and therefore the nearest and farthest hyperfaces are both the same size. Thus, the parallel projection of the hypercube is two cubes joined together to produce a cuboid. As the hypercube rotates, no perspective distortions occur as a result of the projection from four dimensions to three dimensions. A few selected frames from the movie are shown in Figure 4.

Three-dimensional movies of the perspective projection of a four-dimensional simplex, hypertetrahedron, and hypersphere were also generated by the computer. The hypersurface of the hypersphere was specified by randomly scattering points on its surface which were plotted as dots by the computer in the final movie. The points were scattered so as to have a uniform distribution over the surface of the hypersphere.

A five-dimensional hypercube was projected perspectively from five dimensions to four dimensions, and the four-dimensional projection was projected perspectively to three dimensions. However, the final three-dimensional projection, which appeared as a cube-within-a-cube within a cube-within-a-cube was extremely complicated so that the distortions resulting from the rotation were very difficult to follow. Thus, four dimensions would seem to be a practical limit since higher-dimensional objects are presently too detailed to be displayed adequately by the computer.

Discussion

At first it was thought that the computer-generated movies of the four-dimensional hyperobjects might result in some "feeling" or insight for the visualization of a fourth spatial dimension. In particular, perhaps some visualization of a solid four-dimensional hyperobject would be gained from the distortions in the three-dimensional perspective projection. Unfortunately, this did not happen, and we are still as puzzled as the inhabitants of Flatland in attempting to visualize a higher spatial dimension.

However, the importance of the techniques presented

in this paper is the use of a digital computer to generate visual displays of the three-dimensional projections of the hyperobjects. Such displays of rotating hyperobjects could be produced most efficiently by a computer since the projections and drawing would be too tedious and impractical to produce by any other method. Although no actual mental visualization of the fourth dimension resulted from the computer-generated displays, it was at least possible to visually display the projections and be puzzled in attempting to imagine the rigid four-dimensional hyperobject. Of course, these techniques should be useful in displaying data with more than three variables.

The movies have already been useful in extending knowledge of three-dimensional perspective projections to higher dimensions. The techniques have been applied to real-time graphical displays so that the user can rotate, translate, and manipulate hyperobjects and hyperdata and immediately see the results on a graphical display.

Acknowledgments. Grateful acknowledgment is made to Dr. D.E. Eastwood, Dr. M.V. Mathews, Dr. M.R. Schroeder, and Dr. M.M. Sondhi for their lively discussions, enthusiasm, and mathematical assistance in the multidimensioned aspects of this hyperdimensional project.

ACM: Received February 1967; revised April, 1967

References

1. Abbott, Edwin A., *Flatland*. Dover Publications, Inc., New York, 1952.

2. Boerner, Hermann. *Representations of Groups*. North-Holland Publishing Co., Amsterdam, 1963.

3. Coxeter, H.S.M. *Regular Polytopes*. The Macmillan Co., New York, 1963.

4. Kendall, M.G. *A Course In the Geometry of n Dimensions*. Hafner Publishing Co., New York, 1961.

5. Murnaghan, Francis D. *The Unitary and Rotation Groups*. Spartan Books, Washington, D.C., 1962.

6. Noll, A. Michael. Computer-generated three-dimensional movies. *Comput. Autom.* 14, 11 (Nov. 1965), 20-23.

7. Sommerville, D.M.Y. An Introduction to the Geometry of
       N Dimensions. Dover Publications, Inc., New York,
       1958.

8. Stromberg, Gustaf. Space, time, and eternity.
       J. Franklin Inst. 272, 2 (Aug. 1961), 134-144.

Real-Time Computer Graphics Analysis of Figures in Four-Space

Thomas F. Banchoff and Charles M. Strauss

Real-time interactive computer graphics provides the opportunity for a research mathematician
to investigate directly the geometric properties
of curves and surfaces as they undergo transformations in 3- and 4-dimensional space. In this
paper we will describe two films containing
examples which indicate some of the power of this
method and which illustrate the ways in which the
graphics facility aids in the visualization of
complex geometrical relationships. At the end of
the paper, we give a brief technical description
of the equipment that makes this sort of investigation possible and accessible.

The first film is entitled The Hypercube:
Projections and Slicing. The second is Complex
Function Graphs: Squaring and Exponential Functions. The first film examines one of the most
basic and best understood figures in geometry, the
square and its higher dimensional analogues. The
four coordinates (-1,-1), (-1,1), (1,1), (1,-1)
describe the square in a 2-dimensional coordinate
system and the eight coordinates (±1,±1,±1) give
the vertices of its 3-dimensional analogue, the
cube. Although we can represent the square all at
once on the plane surface of a computer graphics
terminal, we can only present pictures of a cube
by projecting to a chosen plane either by straight
on or orthogonal projection or by central projection from a given point. No single view gives the
entire picture, but by rotating the configuration,
or, what is equivalent, by continually changing
the position of the plane to which we project, we
obtain a sequence of pictures which we can then

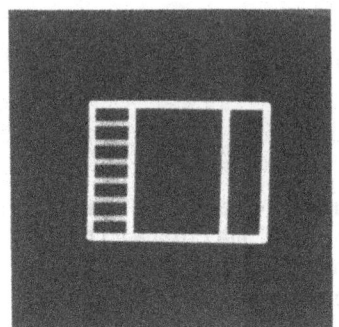

Figure 1

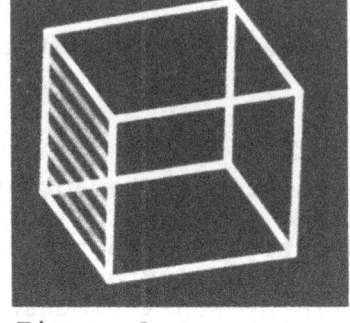

Figure 2

Figure 3

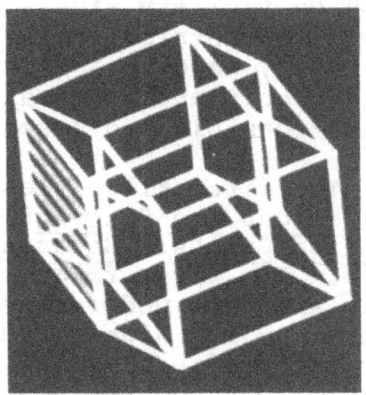

Figure 4

present as a motion picture film. With a machine
that is fast enough to compute many pictures per
second, it is possible to turn a dial or manipulate a "joystick" airplane-type control and have
the figure respond by rotating about various axes
in 3-space, thus producing a collection of images
on the screen which we readily perceive as the
shadows of a revolving cubical framework in 3-
dimensional space.

Certain rotations produce more information
than others, and some rotations are special enough
to produce a high degree of visual ambiguity which
is resolved or clarified only by a subsequent rotation. For example, the straight-on view of a
cube looks like a square and then, as it begins to
rotate about an axis parallel to one edge, like a
collection of squares or rectangles sliding
through one another, an effect well-described as
the "revolving door illusion". (Figure 1) A rotation of 120 degrees about the vertical followed
by 60 degrees about the horizontal produces an
image which contains six parallelograms but which
we readily interpret as the images of the six
identical square faces of a cube. (Figure 2)

All of these ideas extend to four dimensions.
Instead of using pairs or triples of real numbers,
we specify the coordinates of a point by giving
four numbers. The corners of the 4-cube are then
given by $(\pm 1, \pm 1, \pm 1, \pm 1)$ so we get 16 vertices. We
have to describe which pairs of vertices are to be
connected by edges but this is accomplished by
using the same rule that works for the square and
the 3-cube, namely, we join two corners by an edge
if and only if the coordinates are different in
exactly one of the components. Thus we join
$(-1, -1)$ to $(-1, 1)$ but not to $(1, 1)$, we join
$(-1, 1, 1)$ to $(-1, 1, -1)$ but not to $(1, -1, 1)$, and we
join $(-1, 1, -1, -1)$ to $(-1, 1, 1, -1)$ and $(1, 1, -1, -1)$
but not to $(-1, 1, 1, 1)$ or $(1, 1, 1, 1)$. Just as two
edges come from each vertex of the square, we get
three edges from every vertex of the 3-cube and
four from every vertex of the 4-cube. We can use
this fact to count the numbers of vertices in each
case, since for example we have 8 vertices for the
3-cube, each with 3 edges, yielding 24. But this
counts each edge twice, so we end up with 12 edges
for the 3-cube. Similarly we get $(16 \times 4)/2 = 32$

edges for the 4-cube.

Naturally it is possible to analyze these combinatorial configurations in great detail and to proceed to generalize them to arbitrary dimensions, and indeed it is this manipulation of analytic geometry concepts which most often characterizes higher dimensional geometry. For four dimensions however, it is still possible to gain a considerable amount of geometric intuition visually, by interpreting projections of the vertices and edges of the 4-cube into 3-dimensional space. For over a century geometers have drawn pictures and constructed models of such projections, with varying degrees of success in communicating insights into the geometry of an object which cannot be fully described by any model in 3-space. The power of the computer graphics approach is that we can take any projection into 3-space and manipulate it by rotations and projections as if it were a wire-frame model in some room which we were viewing through the window provided by the television screen of the graphics terminal. But in addition to these rotations about an axis in 3-space, we can also turn a dial that effects a rotation in 4-dimensional space and we can watch the effect on the images of the projections into 3-space and then down to the 2-dimensional screen. Occasionally we will project two slightly displaced images which we then observe using stereoscopic viewing apparatus, but in general we achieve enough of a 3-dimensional effect by having the figure rotate slowly and letting the motion clues provide the illusion of depth.

For the 4-cube again there are certain rotations which are difficult to understand fully at first because the figure is not sufficiently unfolded in the 3-space into which it is projected. The first rotation produces the 4-dimensional analogue of the revolving door illusion for the 3-cube, a set of sliding cubes which stop after a 120 degree rotation to resemble the framework of a box kite. (Figure 3) In the film one cube is colored red and the opposite cube is green, so it is easier to fix attention on a particular part of the rotating 4-cube. This is especially useful as we proceed to rotate in a different direction to get a figure in which it is possible to see all 16 vertices that we expect on the 4-cube and all 32 edges. (We have 12 edges each on the red 3-cube and the green 3-cube, and eight more edges joining their corresponding vertices.). (Figure 4)

This figure, rotated in two different ways in 4-space, is still flat in one direction, which we can see if we rotate so that the image of one of the 3-cubes in the boundary is seen to lie in a 2-dimensional plane. A further rotation in 4-space then gives a projection which has a truly general position, with four edges at each vertex no three of which lie in a plane. (Figure 5)

It is possible to view the same sequence of rotations using central projection rather than orthographic, so that we get a perspective effect, with the back face smaller than the face in the foreground. (Figure 6) Rotating one of these central projections of the 4-cube produces some of the best effects in this Hypercube film. (Figure 7)

Another techniqe which is extremely effective in analyzing figures of high complexity is the technique of <u>slicing</u>. We choose a collection of parallel planes and we show what the successive slices are as the planes move across the figure.

In 2-dimensions, we slice a square by a line and we get a family of parallel segments, all of the same size if the slicing line is parallel to an edge and growing and then decreasing segments in the case where the slices are perpendicular to a diagonal. (Figure 8)

In 3-dimensions if we slice a cube square first we get a family of parallel squares, and edge first gives rectangles growing and then decreasing back to an edge.

Most interesting is the set of slices corner first, perpendicular to a long diagonal. An exercise in spatial visualization which confuses many students is to ask what the slice is which occurs half way through. The answer is provided by the picture. (Figure 9) Halfway through we cut each of the six squares on the boundary of the 4-cube exactly the same way and we get a regular hexagon.

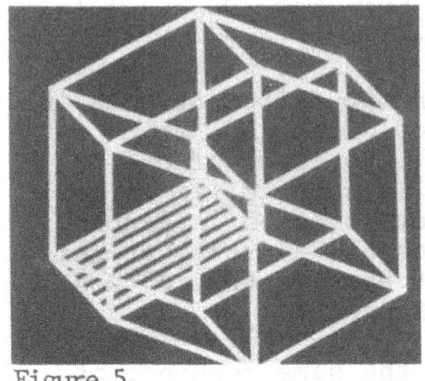

Figure 5

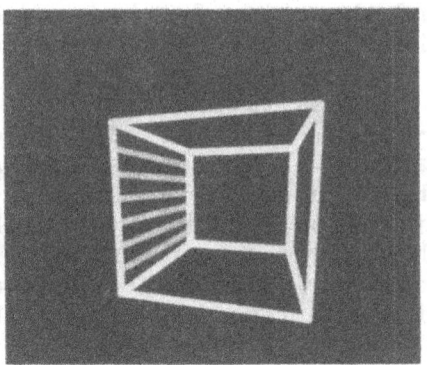

Figure 6

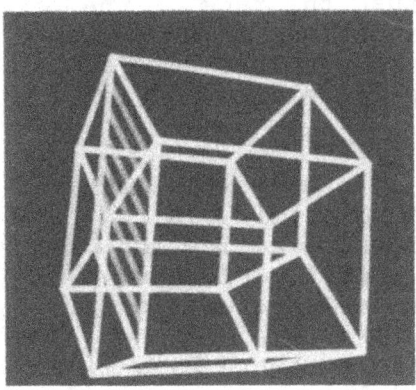

Figure 7

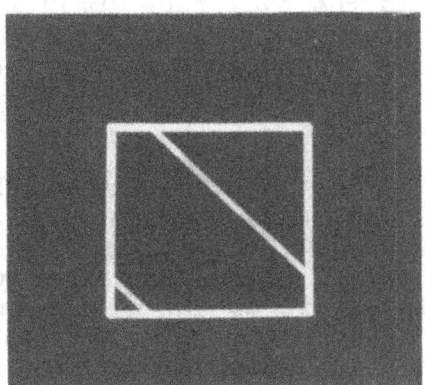

Figure 8

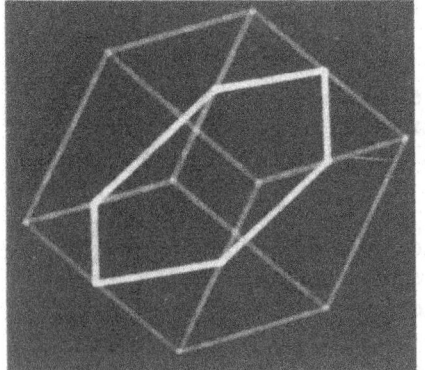

Figure 9

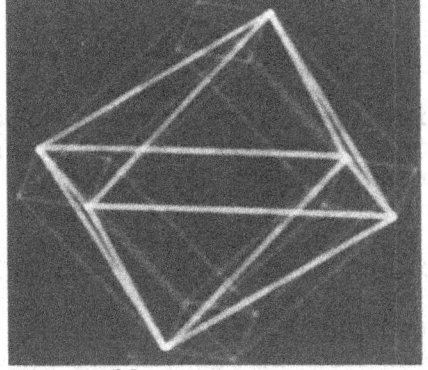

Figure 10

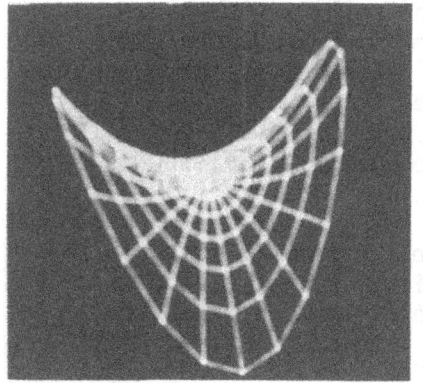

Figure 11

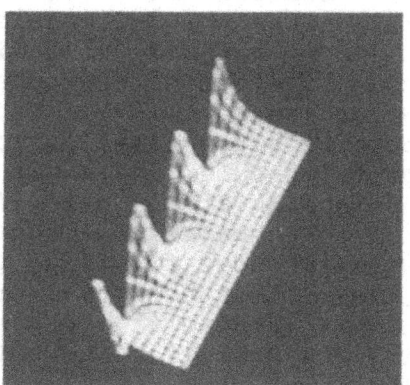

Figure 12

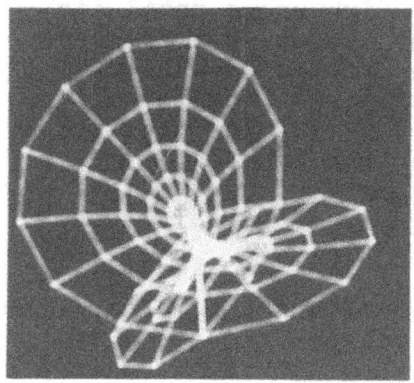

Figure 13

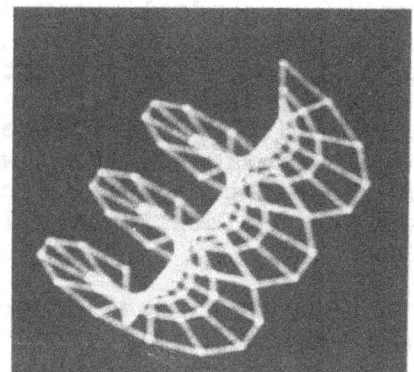

Figure 14

Exactly the same sort of program enables us
to analyze the 3-dimensional slices that are produced by a family of 3-dimensional hyperplanes as
they move across a 4-cube in 4-dimensional space.
We obtain a family of cubes when the slicing hyperplanes are parallel to one of the 3-cube faces
of the 4-cube, and a set of square prisms beginning
and ending at a square if the slicing hyperplanes
are taken parallel to one of the square faces. If
we slice edge first, we get a set of triangular
prisms, then hexagonal prisms, then triangular
prisms, ending at a segment. Again, most interesting are the slices we get corner first. (Figure
10)

The slice begins with a point then moves to a
small tetrahedral pyramid which grows until its
corners become cut off thus producing a truncated
tetrahedron. Three-eighths of the way through the
cube we obtain an Archimedean semi-regular polyhedron with four equilateral triangles and four
regular hexagons. Half way through we meet each
of the eight bounding 3-cubes exactly the same way
and the 3-dimensional slice is a regular octahedron.

The techniques of projections and slicing can
be used especially effectively in the study of
complex function graphs. We write the complex
function $w = f(z)$ in terms of its real and
imaginary parts, $w = u + iv$, $z = x + iy$,
so $u = u(x,y)$, $v = v(x,y)$. We then enter
the graph as a parametrized surface in Euclidean
4-space, $(x,y, u(x,y), v(x,y))$. We can project
into the 3-dimensional subspace $v = 0$ to display
the real part of the complex function, then rotate
in 4-space to transform the image into the imaginary part, in the subspace $u = 0$. Rotating
another way in 4-space, to the subspace $y = 0$,
we transform the graph of the real part of the
function into the graph of the real part of the inverse relation, which we can then rotate to display the transformation from the real to the
imaginary part of the inverse relation.

In the film, we display the function graphs
of $w = z^2$, or $u = x^2 - y^2$, $v = 2xy$ (Figure 11) and of $w = e^z$ or $u = e^x \cos y$,

$v = e^x \sin y$. (Figure 12) In each case the projections provide a visualization of the Riemann
surface of the function, as well as the Riemann
surface of the inverse relations, the square root
(Figure 13) and the logarithm. (Figure 14) This
film of the exponential function has also been
described in [1].

The intersections and singular points of
these projected surfaces provide some of the most
significant information related to the geometry
and topology of surfaces in 4-space. In particular, the origin of the graph of the square root
relation is an example of a "Whitney umbrella",
studied in the theory of normal singularities and
normal characteristic classes. Such applications
to research in geometry and topology are the subject of an address to be given at the International Congress of Mathematicians in Helsinki, August
1978, and details will appear in the Proceedings
of that Congress.

Production Notes--These films were produced
on a Vector General Scope attached to a Digital
Scientific Meta 4 computer, augmented by a parallel processor, the SIMALE, developed at Brown University. Copies of the film for sale or rental
are available from Banchoff/Strauss Productions,
Brown University, Providence, Rhode Island.

References

[1] Banchoff, T. and Strauss, C., Real Time Com-
     puter Graphics Techniques in Geometry, Pro-
     ceedings of Symposia in Applied Mathematics,
     Vol. 20, American Mathematical Society
     (1974), 105-111.

9

Randomness and Order
in Sculptural Form

Harriet E. Brisson

My sculpture is based on an intuitive approach to
form whether working with close packing structures or random
patterns on clay surfaces. My background is in art, not in
mathematics; however, my work has a strong basis in
mathematics from a visual point of view. I understand
relationships through the actual manipulation of tangible
forms rather, than as the result of an abstract or symbolic
study of the equations by which they may be generated.

Recently my sculpture has been an investigation in
two directions which superficially seem to be diametrically
opposed. One direction is the construction of close packing
forms; the other is the production of random patterns on clay
surfaces.

Initially I believed these random patterns were
totally the result of chance relationships, but I have since
realized that they are the result of a set of events which
have as real a mathematical definition as close-packing
structures. The difficulty lies in understanding these
mathematical relationships with the same clarity as that of
the more obviously ordered geometry of the close packing. I
am presently concerned with making this visible in my work
with the "random" pattern on clay surfaces.

Because of the linear nature of boards and steel
beams, close-packings in the form of cubic lattices are a
predominant part of one's living environment and thus are
more readily understood because of their familiarity. In
early childhood one is introduced to this structure by means
of building blocks. In school this is reinforced by studying
the world in three dimensions, once again based on the cubic
lattice. First year architecture students discover, with
amazement, the close-packing of tetrahedra and octahedra

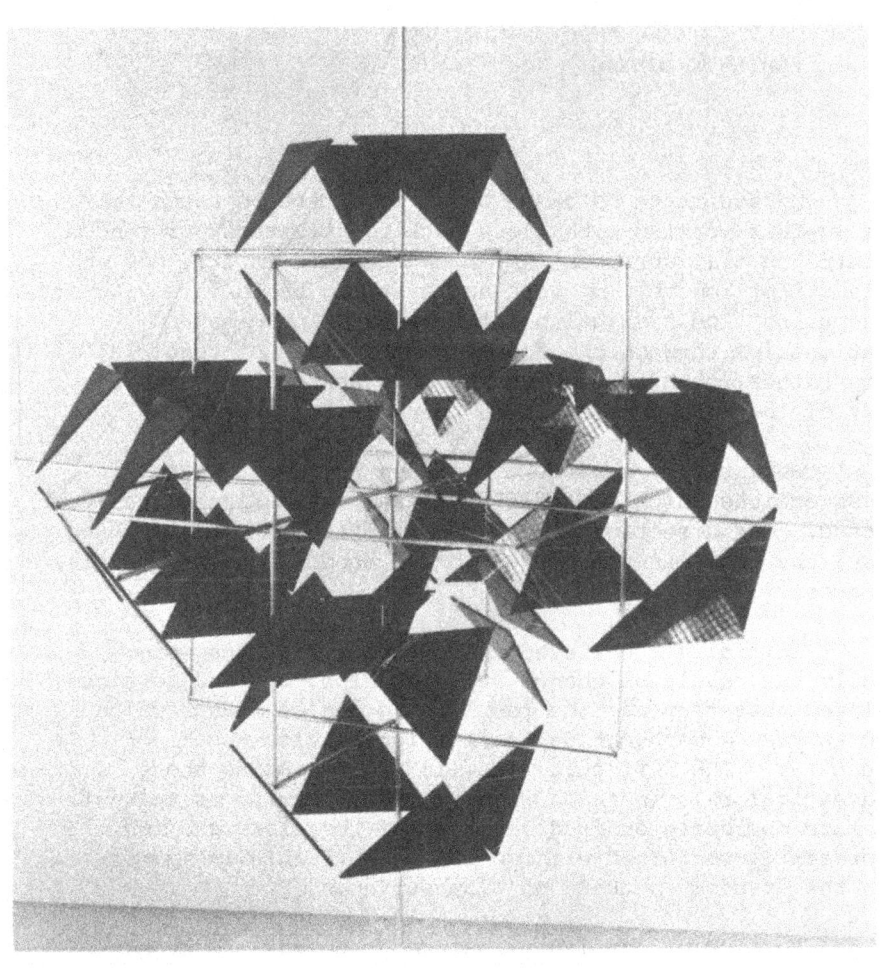

known as the space frame or octet truss, when given the
problem of putting together regular polyhedra that completely
fill space.

My more recent investigation into the visual
expression of order concerns constructing a series of close-
packing tensegrity structures out of aluminum tubes, nylon
cord and plexiglass. This work has given me a useful basic
understanding of the character of structure. Out of this
grew one of my most recent pieces which is an "infinitely
box" containing neon tubes in the shape of a great
rhombicuboctahedron and six octagonal prisms. The single
unit becomes an infinite structure inside the mirrored cube
and is visually evocative as well as producing a sense of
mystery and ambiguity, highly ordered an infinite repetition
of a single unit filling space completely.

Unfortunately this piece could be viewed only through
its single one-way mirrored front surface. To broaden this
experience I built a second neon piece with one-way mirrors
on the four vertical surfaces of the cube, so that it could
be seen from all sides. The form is a truncated octahedron
repeated infinitely by the mirrored surfaces. The mirrored
surfaces appear solid when behind the form, but transparent
in front of the form. Thus it becomes a "Magic Box" which
has sides that are opaque from one view, but dissolves when
the stronger light behind them turns them into "glass"
through which the infinite structure can be seen.

At the same time I have been working with clay, but
in quite a different manner using an intuitive approach. I
have produced random patterns on flat tile tesselations and
and on curved clay forms by covering them with slips
containing different metallic oxides and then subjecting them
to a variety of firing techniques and fuels. The patterns on
the surface show the flow of the flame as it moves across the
form. Sawdust was used as the primary fuel in some; in
others it was the secondary reducing agent after raku
firing in a kiln using charcoal, gas or kerosene for fuel.

This work depended upon the interaction of several
variables: form, surface coating, firing technique and fuel,
which produced random patterns and color variation due to
various phenomena occuring in a specific predetermined
sequence. When a desireable, aesthetically pleasing or
interesting result was obtained, it was necessary to
determine the set of events that had produced that effect.
Once this was achieved and tested, it was possible to
duplicate the results with a high degree of predictability.
Every individual piece in a given sculptural form had exactly

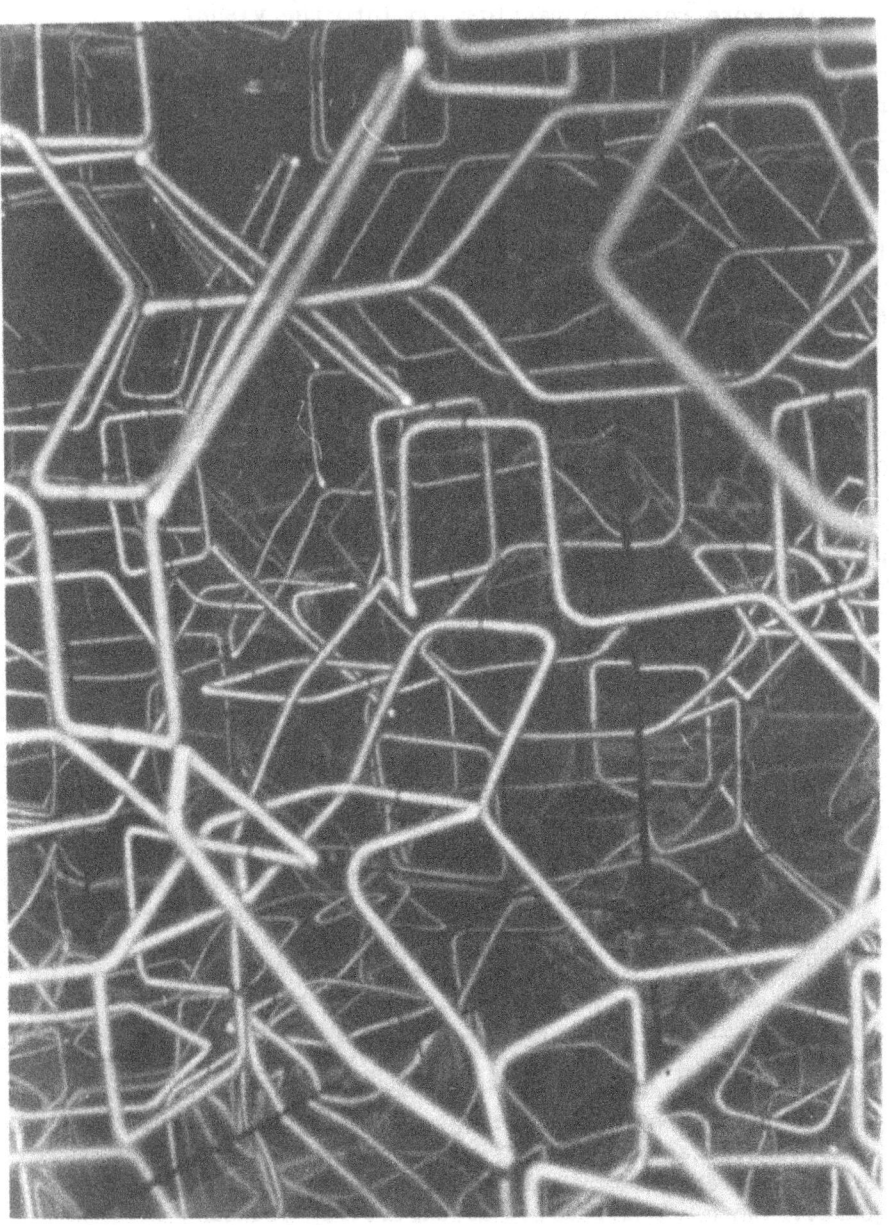

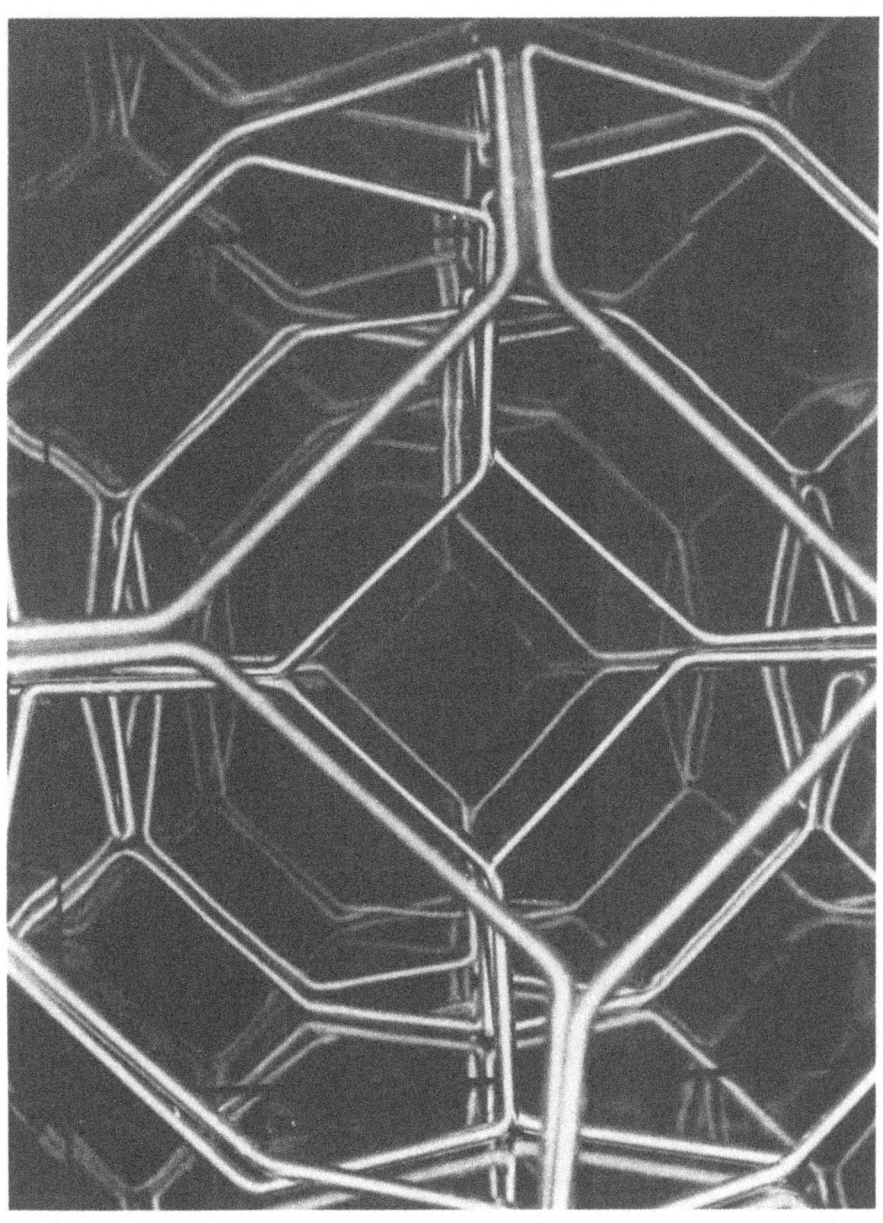

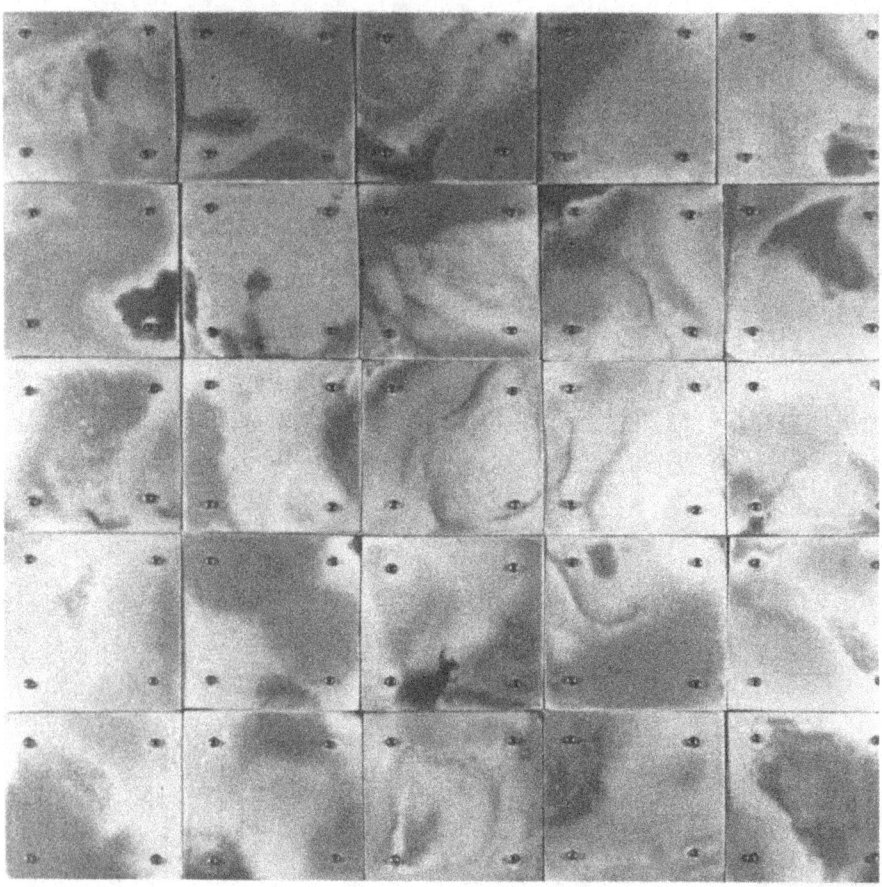

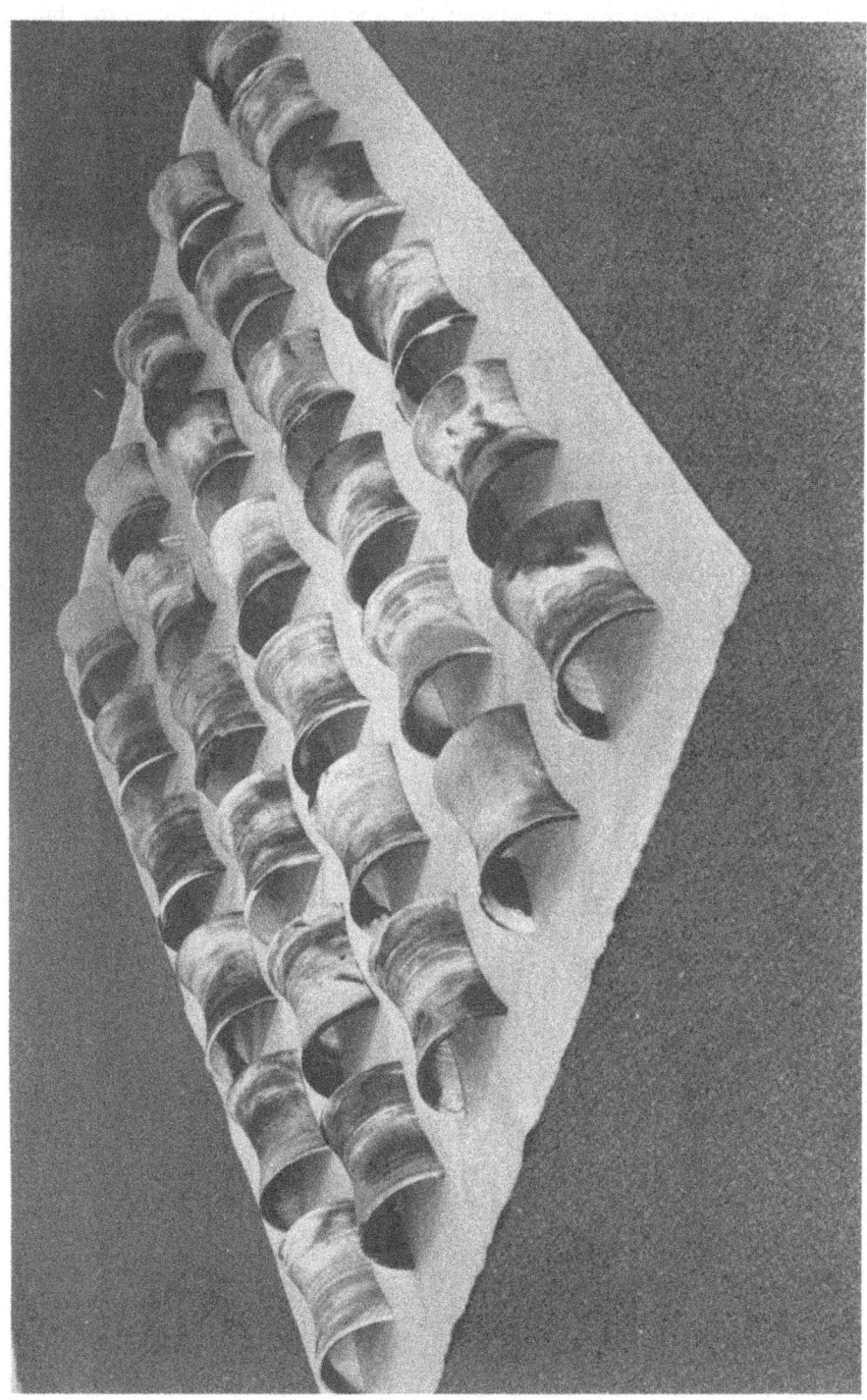

the same combination of materials and firing techniques, but the variation on the surface demonstrates the parameters of this particular set of events.

In this work I have been exploring the meaning of randomness as a visual concept. The surface patterns produced are random; however, the process by which they are produced is not left to pure chance. They are the result of mathematical relations which are just as subject to law as those used to produce close-packing structures, but they are not as easily grasped.

Thus my work, which apparently is moving in two completely opposite directions, is actually following through an investigation of the visualization of mathematical ideas which began with simple orderly structure and has expanded into considering the more complicated concept of randomness.

Finally, the basic quality which perhaps sets this work apart from much sculpture of the past is my concern with transcending sculpture conceived of as an object. The common aesthetic element shared by the work in clay and the close-packings is that both use segments of infinite structures that transcend the classical "object" of sculpture. The tile tesselations can extend infinitely in two directions, the tensegrity close-packings can extend infinitely in three-dimensions and the neon pieces are reflected infinitely by mirrored surfaces. In each case I have given the viewer a sample unit, but the objective is to extend the viewer's concern to infinity.

10

Ambiguous Structures

J. M. Yturralde

In the universe of which we all form a part, we question our own reality through the various paths of knowledge available to us. Fundamentally, these paths are determined by sciences and art. These two disciplines are complementary; each influences the other, often exciting a considerable mutual enrichment.

These disciplines themselves are not isolated quantities about which we can speak in absolute terms; rather, they are the product of diverse investigations of reality which many cultures have established as the primordial key in human behavior.

The work that I have undertaken since 1967 in spatial, formal and color research is contained in the paintings which I call "Serial Structures." This work forms a part of the modern trend of artistic investigation which bases itself in scientific data.

A systematic and precise search for elemental patterns whose basis is the mathematical and geometric structure of sensory reality, has led to the aesthetic for-malization of a number of sequences of such structures in my painting. One possible definition of a structure of any kind, be it paiting, architecture, snowflakes, marine protozoa, etc., is as "a basic support aggregate that forms a significant and physically stable totality, capable of provoking a characteristic emotional response." I am interested in all the aspects of this process, from the formation of the structure itself, to its perception by man and his consequent response to that perception.

This paper is a limited explanation of my own experiments with the expressive modulation of space. One of

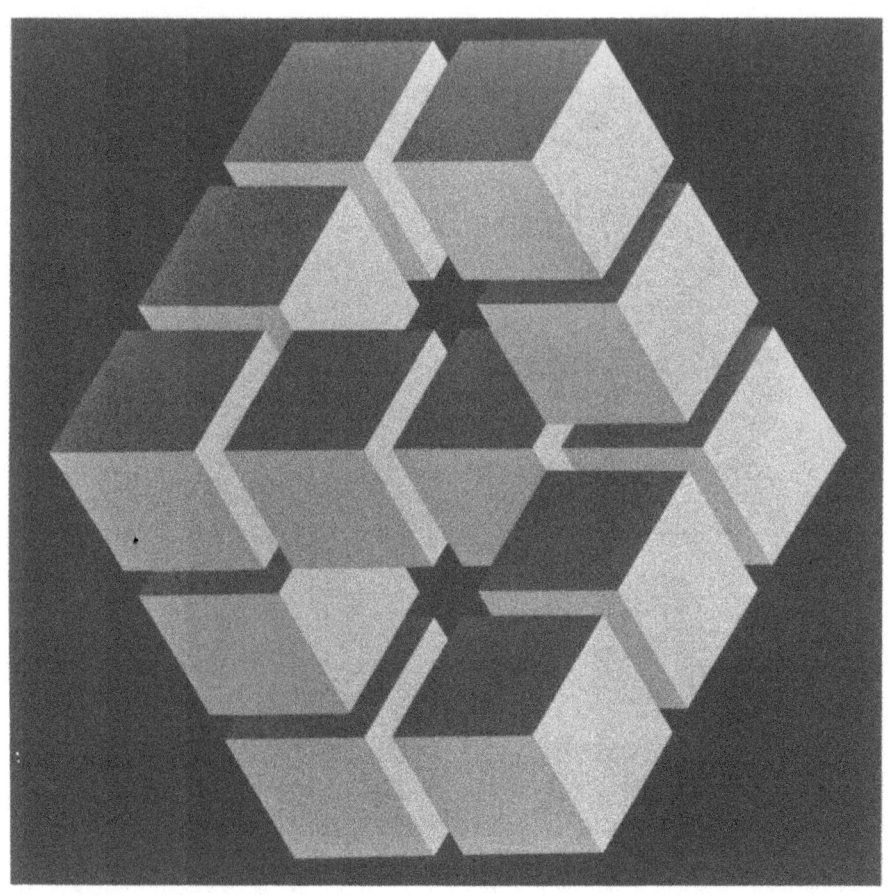

my basic objectives has always been to communicate, through
painting, with clarity and simplicity.

Using linguistic theory as a starting point, I began,
in 1967, to study the possible ways to objectivize aesthetic
information. My interest, since that time, has centered on
the nature and qualities of the data which stimulate specific
perceptual situations.

I have treated the two-dimensional plastic medium as
a language which has its own syntax, which is open to
analysis. The expressive elements of this language--color,
form, texture, etc.--are dictated by a concern for diverse
circumstances: the intended location of the work, the
physical environment, and the social group to which the
painting is directed. All of these elements come together to
produce a customized version of an archetypal image or
pattern. I have been especially interested in the
construction and perception of ambiguous or impossible
figures. Essentially they consist of the representation of
apparently three-dimensional structures with two-dimensional
data which do not provide sufficient information to locate
in depth the different parts of the figures. Our sensory
experience is thus insufficient to clarify the ambiguity of
this information. At the sight of an impossible figure the
usual tendency is to accept it at first as normal and to
realize the impossibility of its existence only when one's
eye runs over the whole figure in an attempt to analyze it.
In this manner there is a conflict between the tendency to
structure perceived data and the analysis of the presented
impossibility.

The figures here presented form a part of several
sequences dealing with the expressive modulation of space
produced by elemental forms.

They all exhibit the following characteristics:

1. An interest in and exploration of the
artist's own creative processes.

2. The attempt to create at the highest possible
artistic level considered, and clearly communicative
images. This implies the treatment of aesthetic content
or data as a form of measurable information.

3. An interest in producing various degrees of
psychological and emotional tension in the viewer.

I have tried to convey the highest possible quantity

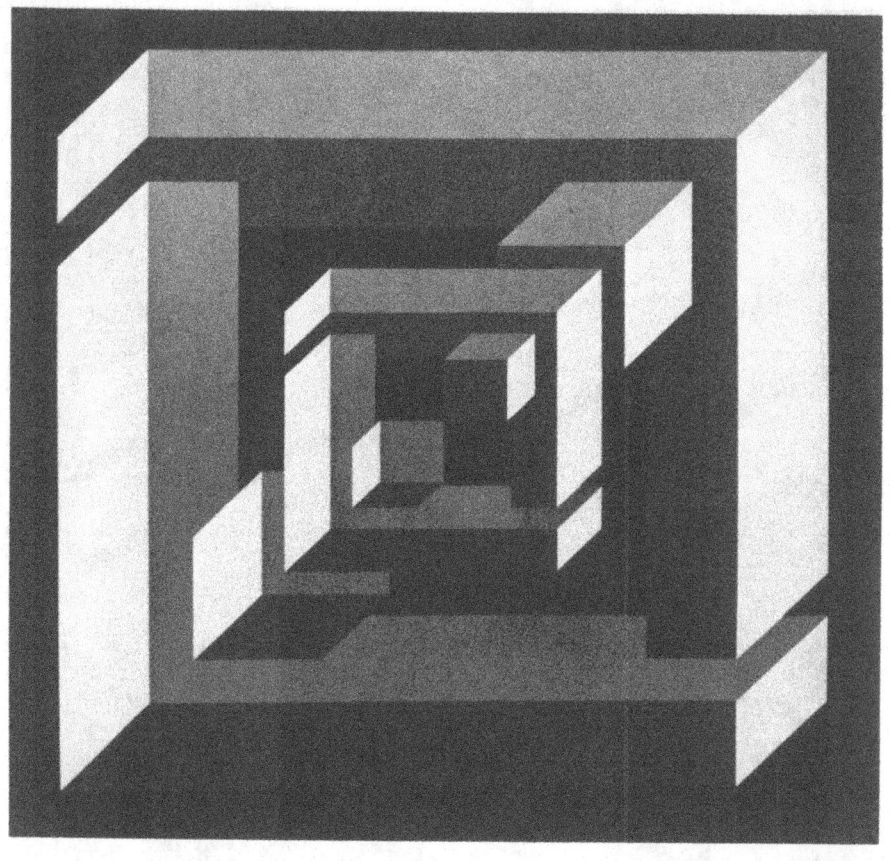

of information about my improbable forms while at the same
time avoiding any kind of pre-established order or continuity
dictated by our visual experience. In this way the painting
seeks a new order, a new dimension of perception, which
should stimulate the act of interpretation in the spectator
and bring about a state of active participation by means of
the mental reconstruction of the suggested forms.

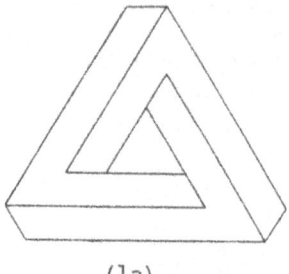 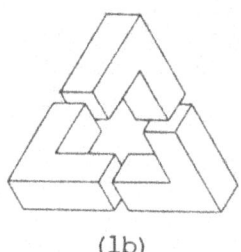

(1a)   (1b)

Penrose's "impossible triangle";
the underlying structure.

(2a)

The phantom gray dots.

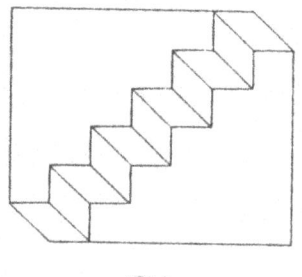

(2b)

The reversible staircase --
an ambiguous figure.

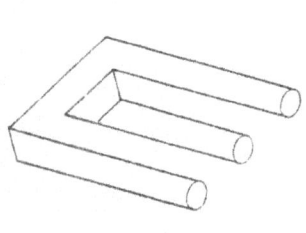 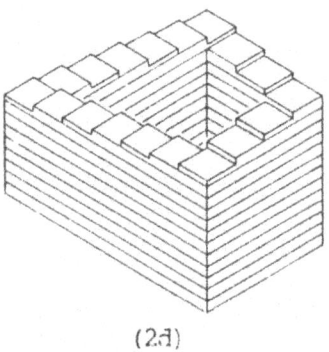

(2c)   (2d)

The tuning fork --   The endlessly rising
a figure-ground   staircase -- a depth
impossible object.   inconsistent impossible
                           object.

11

An Impossible Four-Dimensional Illusion

Scott E. Kim

Introduction

Penrose

In 1958, L. S. and Roger Penrose introduced a new optical illusion called the "impossible triangle" (figure 1a, [Penrose 1958]). The impossible triangle appears to be a 3-dimensional (3-d) structure composed of rectangular beams. But as with all optical illusions, appearances can be deceiving — no consistent 3-d model of the impossible triangle can be built. It exists only on paper. The Penrose article included several variations on the impossible triangle, one of which is shown in figure 2d (the endlessly rising staircase). The artist M. C. Escher has presented a striking variant on the illusion, shown in figure 55a ("Waterfall").

The underlying structure of the impossible triangle is a circuit of three right angles (figure 1b). Each corner individually is possible. If we join two of the corners, leaving out the third, the resulting structure appears to twist 3-dimensionally in a perfectly reasonable manner. If we continue the two free ends, the extended beams should miss each other. In this picture, however, they connect, completing the circuit. Penrose takes advantage of the ambiguities in representing 3 dimensions on 2-d paper to create a false connection.

But there's more to the presentation of the impossible triangle than just a triad of right angles. Figure 1a does not use any perspective — beams do not shrink as they recede. In the original drawing of the impossible triangle, the 3-dimensionality was emphasized by using perspective foreshortening. This necessitated using beams with rectangular rather than square cross sections. Furthermore, the various surfaces were shaded to suggest a light source.

Premise

The impossible triangle is a 2-d drawing which appears paradoxical to a 3-d person. In this article, we will construct the 4-d analog of the impossible triangle. The result will be a 3-d "drawing" which would appear paradoxical to a hypothetical 4-d person. This 4-d optical illusion was first developed by R. Penrose several years after he developed the original impossible triangle. The present article recounts the intuitions that went into my own rediscovery of the idea.

Admittedly, this is a rather peculiar business: combining optical illusions and 4-d space as if one topic weren't confusing enough. Penrose has written that

> "I had originally made a model from cardboard nearly 20 years ago, in what I had always considered to be a 'fit of madness' — since only a 4-d being could properly appreciate it!" [Penrose 1976]

But precisely because the problem is elusive, we are obliged to sharpen our understanding of 4-d space and visual perception. Most of this article is devoted to the spatial and visual intuitions necessary for building and appreciating a 4-d optical illusion. If you think you already have these intuitions, you are encouraged to try the construction for yourself before going on.

This article is structured as a strict palindrome. The first half proceeds from general to specific, starting with the 3-d impossible triangle, deconstructing it element by element, finally reaching an analysis of 3-d vision itself. The second half reverses the procedure, proceeding from specific to general, starting with the elements of 4-d vision, eventually constructing the 4-d analog of the impossible triangle. The article ends with an invitation to try generalizing other illusions to 4 dimensions. Making comparisons between the analogous sections in the first and second halves is useful; much of the information does not lie in either individual half, but in the comparison of the two.

Preparatory Remarks

This article assumes a basic understanding of 4-space — the notion of 4 mutually perpendicular spatial axes. Familiarity with basic linear algebra is helpful but not necessary. The ability to visualize 3-d objects is far more important, since the emphasis in on illustrations, not numerical proofs. Many of the observations in the 3-d half of the article may seem self-evident, but remember that they will reappear one dimension higher in the 4-d half where they may not be so evident.

You may find it useful to make models; a rhombic dodecahedron and several rhombohedra would be particularly useful. Drawings of 4-d objects done on 2-d paper are necessarily very misleading. We are used to pulling 3 dimensions out of 2, but that is not enough here. Ideally, we should represent 4-d objects by actual 3-d models. Unfortunately, this article is only printed on 2-d paper.

Therefore, an illustration such as figure 34a (the isometric projection of the 4-cube) is to be interpreted not as a 2-d image to be seen as a 3-d object, but as a 3-d image, here represented in 2 dimensions, to be seen as a 4-d object. Most of the time, the projection from 3 to 2 dimensions will be chosen as asymmetrically as possible, taking advantage of ordinary 3-d perspective to bring out the 3-d shape. Projections from 4 to 3 dimensions will be chosen as symmetrically as possible to keep things simple, and will usually not involve (4-d) perspective.

Throughout this article, a number appended to the beginning of a term will give the absolute dimensionality of the object. Thus, "cube", "3-cube", "3-d cube" and "3-face of a 4-cube" all refer to the same object. The dimensionality of an illusion, however, will be determined by the dimensionality of the observer, not that of the drawing itself. Since the impossible triangle makes sense only to a 3-d viewer, it will be classed as a 3-d illusion, even though the drawing itself is only 2-d.

The Impossible Triangle

In what sense is the impossible triangle really impossible? We can better understand the source of the paradox by putting the impossible triangle in the context of other optical illusions. (See [Gregory] for further discussion of optical illusions and the intricacies of visual perception.)

Different Types of Illusions

An optical illusion is a visual pun. It offers one interpretation, then just as quickly denies it with another. Puns can occur only where there is ambiguity. A

verbal pun, or "play on words", takes advantage of words
which have dual meanings. Similarly, a visual pun, or
"play on vision", takes advantage of images which have dual
interpretations.

Verbal punning is possible because the process of
understanding language is not consistent. A listener
translating a spoken sound into an internal meaning must
choose among many different mappings. Here, the problem
clearly lies in the structure of the language. Different
languages have different inconsistencies; thus puns do not
usually translate well from one language to another.

With visual puns, there is a tendency to say that
something is wrong with the picture, rather than blame the
process of visual perception. Seeing is believing. But
vision, like spoken language, is a learned skill, full of
arbitrary rules and assumptions. Once these assumptions
have been isolated, it is easy to discover the possible
inconsistencies. Puns do not create ambiguity; they merely
reveal ambiguity which is already present in the language.

Ambiguity alone, however, does not make a pun. Skill
is required. After all, our visual world is constantly
plagued with small ambiguities, but most are easily
resolved. An effective illusion must present a large
discrepancy in the most direct possible way. Furthermore,
the paradox must not be easily resolved; the two
interpretations the illusion presents must be truly
irreconcilable. Ideally the two interpretations should be
reached by separate means. In the following examples of
optical illusions, observe how an ambiguity of visual
perception has been exploited.

In figure 2a, gray dots seem to appear at the
intersections of the white streets. Dots only appear the
intersections at which you are not looking. They disappear
if looked at directly. This illusion is based on the fact
that color perception is heavily influenced by edge
constrasts. The effect is made noticeable by the stark
geometric pattern.

Figure 2b is an example of a "multi-stable figure"
[Attneave]. This picture can be interpreted as a
legitimate 3-dimensional structure (assuming that an
infinitely thin back wall is acceptable). Unfortunately,
it has another interpretation which reverses all the folds,
equally valid, but inconsistent with the first. (If this
is hard to see, try turning the picture upside-down.) This
ambiguity is based on the fact that a line drawing of a

fold can always be interpreted as either in or out. The inconsistency is brought out by making the figure entirely symmetric — neither interpretation is given priority.

Figures 2c and 2d fall into the category of "impossible objects". Both appear to be 3-dimensional objects, but contain inconsistencies which prevent actual models from being built. The "tuning fork" (figure 2c) relies on the ability of a single line to function either as an upper or a lower edge. The "endlessly rising staircase" (figure 2d), relies on the ambiguity of portraying 3-dimensional depth on 2-dimensional paper. The staircase illusion is closely related to the impossible triangle, which is also classed as an impossible object.

Impossible objects maintain their ambiguity by presenting separate local and global interpretations. Locally, two beams seem to meet at each corner of the impossible triangle, one crossing in front of the other. For instance, the beam entering the top corner from the left in figure 1b seems to cross in front of the beam leaving the corner to the right. If we start at one corner and trace the triangle clockwise, we should keep getting farther and farther away. Globally, the impossible triangle appears to be a single connected circuit. If we start at one corner and follow the circuit clockwise, we eventually return to the same point. By maintaining consistency and inconsistency on separate conceptual levels, the impossible triangle is able to suggest 3-dimensionality without adhering to it.

Impossibility Proof

No consistent model of the impossible triangle can be made, but it is difficult to say precisely why. If we assume that the all corners are right angles, then the angles of the impossible triangle add up to 270 degrees — an impossibility. This assumption is unnecessary, though. The impossibility can be proved without assuming any conditions on the angles. Another approach emphasizes twistedness. Tracing one of the faces of the impossible triangle as it travels around the triangle, we cover the entire surface after four laps. The impossible triangle returns to itself twisted, in the manner of a Moebius strip. We might expect that all possible loops of square beams must be untwisted. The object portrayed in figure 3a, however, is perfectly possible despite its twist.

The following ingenious impossibility proof is due to [Huffman]. Suppose for the moment that figure 3b has a

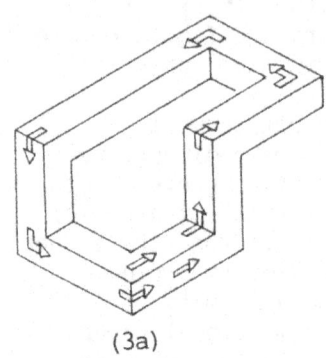

(3a)

A twisted but possible object.

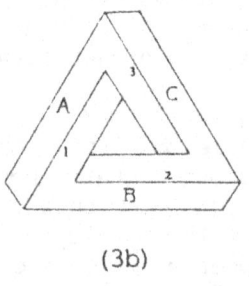

(3b)

The impossible triangle labeled for an impossibility proof.

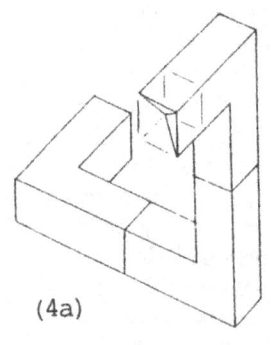

(4a)

Broken.

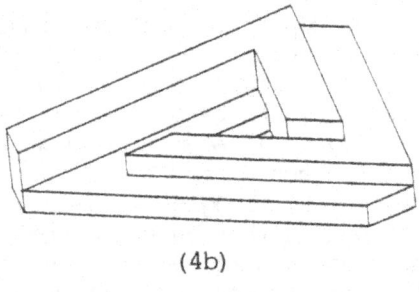

(4b)

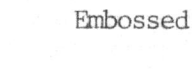

Embossed.

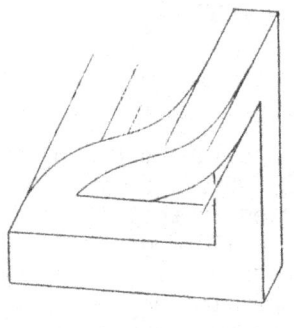

(4c)

Curved.

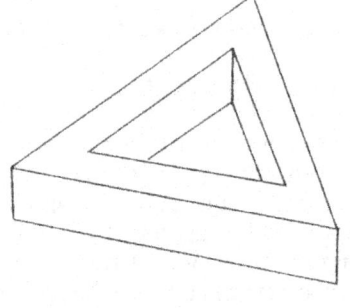

(4d)

Non-illusionary.

False models of the impossible triangle.

valid 3-d representation. What can we say about its
structure? Three faces are visible, labeled A,B,C. We do
not know their exact shapes, but we do assume that they are
planar. Looking at the way surfaces meet one another, we
can see that

> face A meets face B at edge 1
> face B meets face C at edge 2
> face C meets face A at edge 3

No two of the planes containing any of the visible faces
may be coplanar without the model going flat, hence they
are all distinct. But three distinct non-parallel planes
always intersect in a single point (assuming they lie in
the same 3-space). Furthermore, each of the lines
determined as the intersection of two of the planes must
pass through this point. In fact, we can see that the
lines containing edges 1,2,3 do not intersect in a single
point. Therefore, figure 3b does not represent a possible
object.

Models

Where does the impossible triangle go wrong? What
assumption does it violate? If we suppose that each corner
of the impossible triangle individually is valid, as in
figure 1b, then the assumption violated is one of
connectedness — the sequence of beams could not possibly
form a closed loop. Actually, there are many assumptions
made in the impossibility proof, any one of which could be
responsible for the paradox. Let us study some of the
false models which "almost" explain the impossible triangle.

Broken One way to make the model appear to be
connected is to "trim" the overlapping ends as in figure
4a. Ends which are actually at different depths seem to
join smoothly. This model must be viewed from a particular
angle in order for the specially shaped notch in the beam
in front to match the shape of the beam in back. A slight
change in position destroys the illusion. This model
violates our assumption that the model is connected.

Embossed The model in figure 4b does not violate
connectedness. Here, a picture of the impossible triangle
has been "embossed" — each visible region raised to a
different height above the paper. As in the previous
broken model, surfaces which appear to meet at an edge are
actually at different levels, but this time the parts are
connected by invisible surfaces seen edge-on by the
viewer. While the model is connected, it still violates

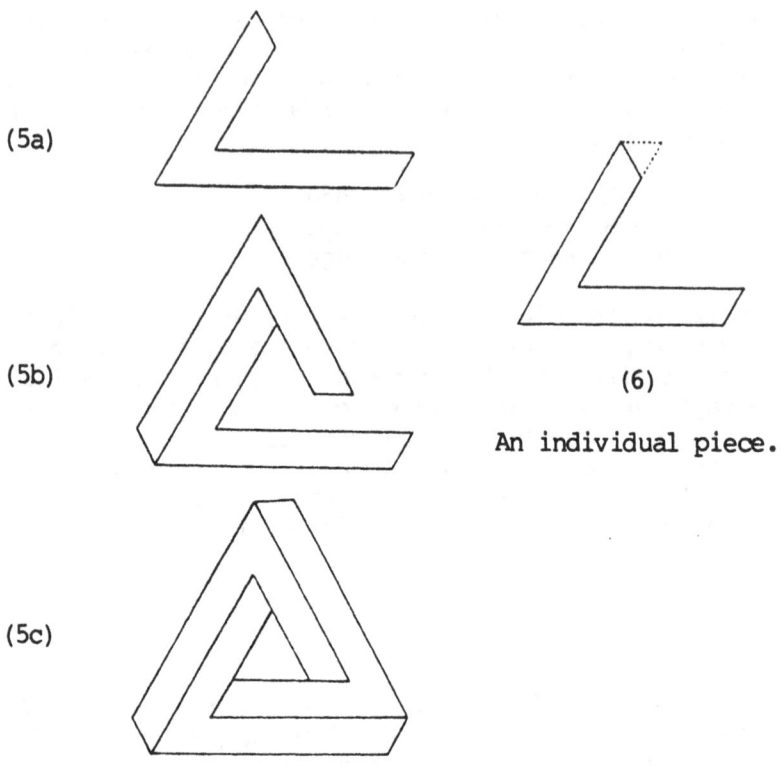

(5a)

(5b)

(5c)

Piece-by-piece assembly
of the 2-d model.

(6)

An individual piece.

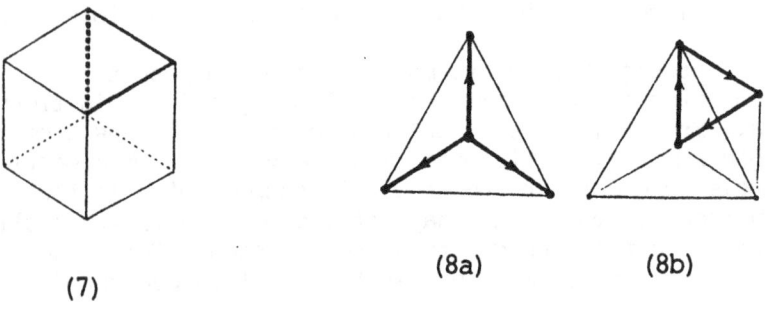

(7)

The triangle extracted from
a projection of the cube.

(8a)   (8b)

Triangle =
rearranged basis vectors.

the "edge-to-edge" assumption — two surfaces which appear tomeet at an edge should actually join each other.

<u>Curved</u> A 3-d model which does not violate the edge-to-edge assumption is shown in figure 4c. The broken end in front has been bent back directly away from the viewer to join the broken end in back. Some of the visible edges are now curved, but if viewed from the correct angle, they will appear to be straight. This model violates the assumption of planarity — individidual surfaces bounded by straight lines in the picture should be planar.

The broken, embossed and curved false models all violate the more fundamental assumption of non-degeneracy — the model must not depend on a particular point of view for its effect. This does not mean that degenerate points of view never occur when observing real objects, only that degenerate models do not explain away optical illusions since they invoke overly complicated assumptions.

<u>Nonparadoxical</u> A very peculiar sort of false model is shown in figure 4d. The triangle portrayed here appears to be perfectly ordinary. But under unordinary circumstances, the type of circumstances we have assumed in the previous false models, we can argue that each part of the model is actually traveling away from the viewer as the circuit is traced clockwise. For instance, what appear to be straight edges might actually be bent away from the viewer as in figure 4c. Under this assumption, figure 4d is an impossible object. There is too little information, however, for this illusion to be convincing. This model violates the assumption of simplicity — the paradox should not be resolvable by a simpler explanation.

<u>Flat</u> There remains one final false model which we have accepted without question all along — the picture itself (figure 1a). This model violates several assumptions. First of all the model is flat — the three faces all lie in the same plane. Our formal impossibility proof breaks down if the model is allowed to go flat. More seriously, the "edges" in the model are not edges at all, but lines drawn on a page (or more precisely, patterns of light and dark which are perceived as lines). This model reveals an ambiguity in the term "impossible triangle". Does "impossible triangle" refer to the 2-d picture, or to the hypothetical 3-d object it purports to portray?

Let us leave 3-space and treat figure 1a as a purely 2-d structure. Figures 5abc show how the impossible triangle can be built from three congruent pieces. The

arrangement is symmetrical — mapping one piece into
another rotates the structure into itself. Each piece can
be described as a large "V", with the two arms running
along two of the three sides of the triangle.

A 2-d person wishing to understand the impossible
triangle might very well make such a 3-piece model, since
only a model which comes apart allows a view of what
happens on the hidden internal surfaces. The 2-person
might also want to color each piece a solid color to
indicate a light source somewhere off in hypothetical
3-space. Note that each piece is composed of a series of
rhombuses corresponding to square surfaces on the original
square beams (figure 6). The one irregular end is clipped
when one beam passes in front of another in 3-space.

Elements of Construction

We have shown how the impossible triangle is
impossible. To produce its 4-d analog, however, it will be
necessary to take a more constructive approach. Assume
that the impossible triangle is a picture of an actual 3-d
object, ignoring for the moment the global inconsistency.
Make the simplest possible assumptions about the
interpretation of the picture. What are the elements which
make up its construction?

Global Form

Why is the global shape of the impossible triangle a
triangle? One way to derive the shape is to rotate a cube
until two opposite vertices appear to coincide. Figure 7
shows such a cube, with three edges darkened. Imagine the
3-d structure formed by this sequence of edges. The point
of view has been chosen so that two opposite vertices of
the cube appear to coincide in the center. As a result,
the sequence of darkened edges appears to form a closed
loop, when in fact the two ends are at different depths in
3-space. Furthermore, the angle of the spurious corner is
exactly like the other two legitimate corners — the loop
is an equiangular triangle. A devious artist could retouch
this corner to imitate the other two, concealing the false
connection.

Another way to derive the shape is to assign
coordinates to the vertices. If the beginning vertex in
back is $(0,0,0)$ and the ending vertex in front is $(1,1,1)$,
then the point of view looks directly down the line $x=y=z$
towards the origin. The paradox is that having traveled a
unit distance in turn along the x, y and z directions you

return to the origin, not (1,1,1).

The global form of the impossible triangle, then, is constructed by rearranging the unit vectors in the x,y,z directions to form a closed loop. We have chosen our projection so that the three basis vectors radiate symmetrically out to the corners of an equilateral triangle. Rearranging these vectors, we obtain another, slightly smaller, equilateral triangle (figures 8ab).

Corners

The most important component of the impossible triangle is the corner. The way corners twist over themselves is responsible for 3-dimensionality, the effect of continual recession. They are also hard to draw. Consequently, the corner needs careful study.

A corner is created whenever two square beams intersect in a cube. Alternately, we can construct a corner by extending a cube in two directions to create beams. Figure 9abcd show the four types of corners which can be pulled from the cube pictured in figure 9e. The black dots and broken lines outline vertices and edges of the corner cube which have been removed by the extended beams. It is useful to superimpose figure 9e on each of the illustrations to get a getter idea of how edges enter and leave the corner.

Note that corner 9b can be considered the "back" of corner 9a. Similarly, corner 9d reverses the relation of hidden and unhidden lines from corner 9b. We can even consider that a surface which folds around the back in one corner is continued in the sibling corner. Corners 9b and 9d appear less natural than other two, since they are seen from underneath.

As 3-d structures, the four corners can be explained as four different views of a single right angle. As 2-d pictures, however, they differ in which faces are visible. The distribution of visible faces, in turn, determines how the corner appears to twist. Let us study the four types of 3-d corners as purely 2-d structures.

Each corner has a different assortment of rhombic "caps". Figure 9c has a cap at either end, while figure 9d has none. Both figures 9a and 9b have one cap. The presence of a cap means that we are able to see the end of the beam, i.e. that the beam is approaching. The absence of a cap indicates a receding beam. All combinations are

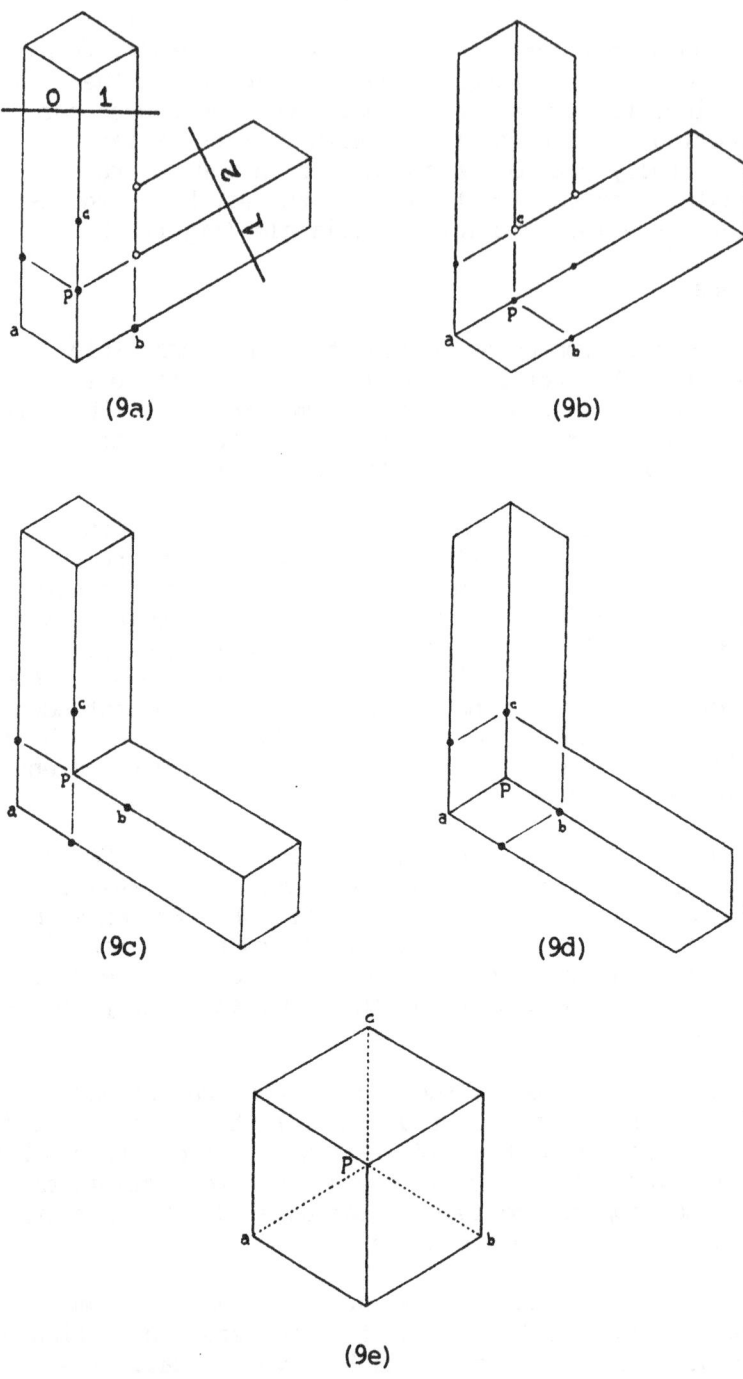

(9a)

(9b)

(9c)

(9d)

(9e)

The 4 types of 3-d corners.

represented here. Figures 9b and 9d are either
"all-positive" or "all negative", while figures 9a and 9b
combine both approaching and receding beams.

We may also study the way the visible surfaces are
redistributed as they pass around a corner. Such an
analysis appears in [Cowan]. Coming into a corner there
are four "surface bands", corresponding to the four sides
of the square cross section. Only two bands are visible at
a time, the opposite two bands being hidden on the back.
If we think of the corner as a racetrack, then there are
two tracks, the inside track having the tighter turn. At a
corner, there are two possibilities. Either the bands will
retain their ordering, or they will undergo a cyclic
permutation in which one band vanishes and a new one
appears.

Suppose we traverse corner 9c starting from the lower
rhombic cap, ending at the upper. As we enter the corner,
we bend gently 60 degrees to the right. The band on the
inside (right-hand) track encounters a fold line (emanating
from point P), but retains its inside position after being
"refracted". The outside (left-hand) band angles off to
the right without folding. The surfaces leave in the same
order they entered. Corner 9d acts similarly. The only
difference is that the outside track is now the one
refracted.

Suppose we traverse corner 9a starting from the
(invisible) right rhombic cap, ending at the upper. As we
enter the corner, we bend sharply 120 degrees to the
right. The band on the inside (right-hand) track vanishes
from sight abruptly at the corner. The band on the outside
(left-hand) track switches suddenly to the inside
(right-hand) track as it turns the corner, cutting off a
triangular tip of the inside band. If we were to add all
such triangular tips to the corner hexagon, we would get a
Star of David (stellated hexagon). As the inside band is
being cut off, a new band appears on the outside track to
maintain the quota of two visible bands.

Described 3-dimensionally, we can say that the beam
twists 90 degrees clockwise as it passes around the
corner. The appearing band signifies that a surface has
folded around from the back side to the front where it can
now be seen, while the disappearing band signifies that a
surfaces has folded around from the front to the back where
it can no longer be seen. The disappearing and appearing
bands occur on opposite surfaces of the square beams, and
so cannot be seen simultaneously.

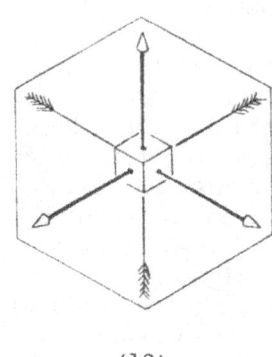

(10)

The six directions
for extending a cube
into a beam.

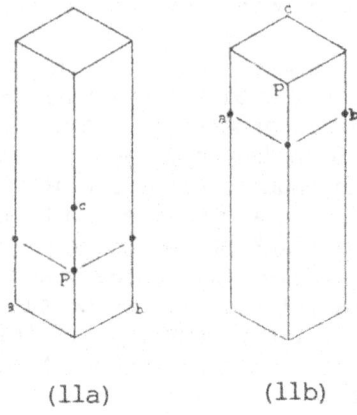

(11a)   (11b)

The two ways of
extending a cube
into a beam.

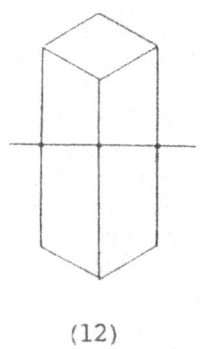

(12)

Sliced square beam.

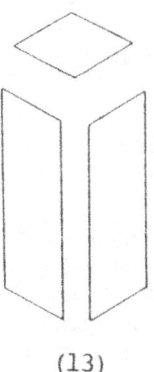

(13)

Exploded square beam.

Since the beam leaving the corner passes in front of
the beam entering the corner, we can say that figure 9a
moves continually forward in 3-space. Corner 9b acts
similarly, except that the fold moves continually away from
the viewer. Each corner of the impossible triangle seems
to move farther away from the viewer. Therefore all
corners are of the type shown in figure 9b.

The twisting can be seen more readily if we assign
labels to the bands. In figure 9c and figure 9d, all
surfaces retain their order, as already described. In
figure 9a, we see that the entering beam has two visible
surfaces, numbered (n),(n+1) left to right (modulo 4).
This property is retained in the exiting beam.

To draw a corner, start with a picture of a cube.
Extending the cube in two directions will produce a
corner. Initially, there are six ways to extend a cube,
one for each of its six faces (figure 10). Extending one
of the three visible front faces produces an approaching
beam, whose front face is visible as a rhombic "cap"
(figure 11a). Note that the top face of the cube (with
opposite vertices P and c) is no longer visible, obscured
by the mass of the approaching beam. Extending one of the
three invisible back faces produces a receding beam whose
cap is invisible (figure 11b). For the impossible
triangle, the corner should combine an approaching and a
receding beam. Be careful not to choose the two beams
along the same axis, or you will get a straight 180° angle.

Finally, superimpose the two extensions on a single
cube. All that is left is to erase hidden lines. These
include edges of the receding beam obscured by the
approaching beam, and edges which are smoothed out when
coplanar surfaces meet.

Beams

The impossible triangle appears to be made of square
beams. A square beam is formed by taking a square and
translating it along a perpendicular axis. This sounds
very much like the definition of a cube, and indeed, a cube
is a very short square beam. Note that the "square" in
"square beam" gives the shape of the cross-section, as it
does in the term "square prism". The word "beam" is used
instead of "prism" only because it has fewer dimensional
connotations.

Figure 12 shows a picture of a square beam with hidden

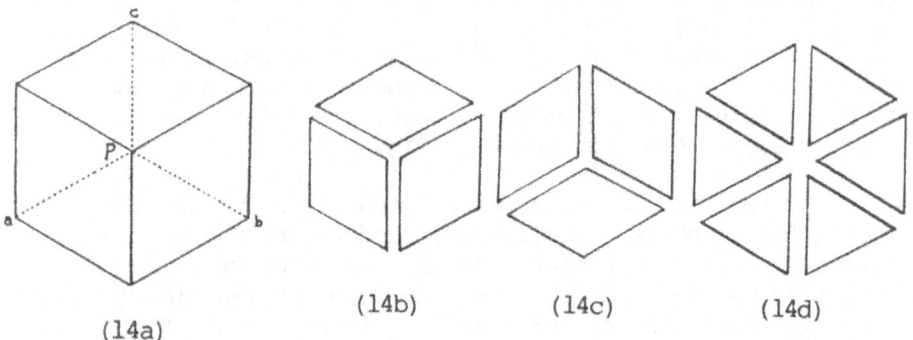

(14a)   (14b)   (14c)   (14d)

The isometric projection
of the cube;

front and back faces.

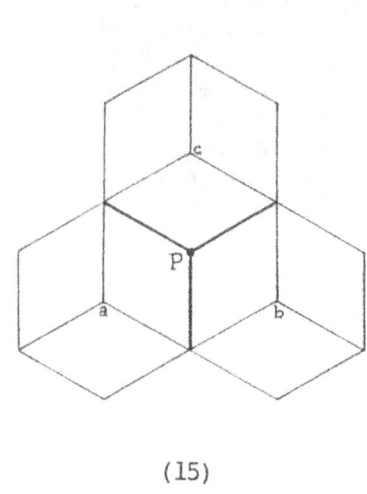

(15)

Excerpt from an
ambiguous rhombic
tiling pattern.

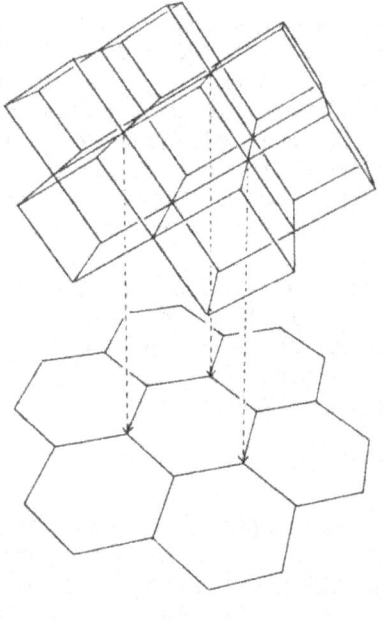

(16)

Tiling a plane
with hexagons.

lines removed. The horizontal line cutting through the
center emphasizes that this is a strictly 2-d figure. Note
that the cross-section of the projection of the square beam
looks like 3 dots — a picture of a square as seen by a 2-d
person.

The three visible faces of the beam decompose the long
column into three parallelograms — two extending down its
length, and one capping the remaining concavity (figure
13). If we were to shrink the length of the beam to zero,
all that would remain would be the cap.

Alternatively, we could start with the cap, and mold a
copy of the undersurface (two line segments meeting at an
obtuse angle, i.e. half a rhombus). This concave shell
will nest smoothly under the cap. Take three rubber bands
and have them join each of the three vertices of the
undersurface of the cap to the three vertices of the
shell. Then an image of a beam (as well as an inadvertent
sling shot) is generated by pulling the shell away from the
cap. Note that this construction is strictly 2-d — no
explicit reference is made to 3-space.

The 3-d Cube

All elements of the impossible triangle are ultimately
based on properties of the cube. A 3-d person wishing to
draw an impossible triangle would find it convenient to use
hexagonal graph paper, which is based on a projection of a
cubical grid. What are the properties of the cube, and how
are they illustrated in its 2-d projection?

The Isometric Projection

Figure 14a shows the most symmetric projection of the
cube into the plane. All edges in this projection map onto
segments of the same length. Thus the projection is called
the "isometric" projection — "iso" for "same", and
"metric" for "measure". In the isometric projection, the
three coordinate axes of space are evenly distributed
around a circle, radiating out to the corners of an
equilateral triangle abc. This extreme 2-d uniformity
simplifies the processing of drawing, but it can also
confuse the eye with unexpected symmetry.

Seen as a 2-d structure, figure 14a is a regular
hexagon with two sets of three lines each radiating out
from the center to alternate vertices. The eight vertices
of the original cube project in two different ways. Six of
the vertices form the hexagon, while the remaining two

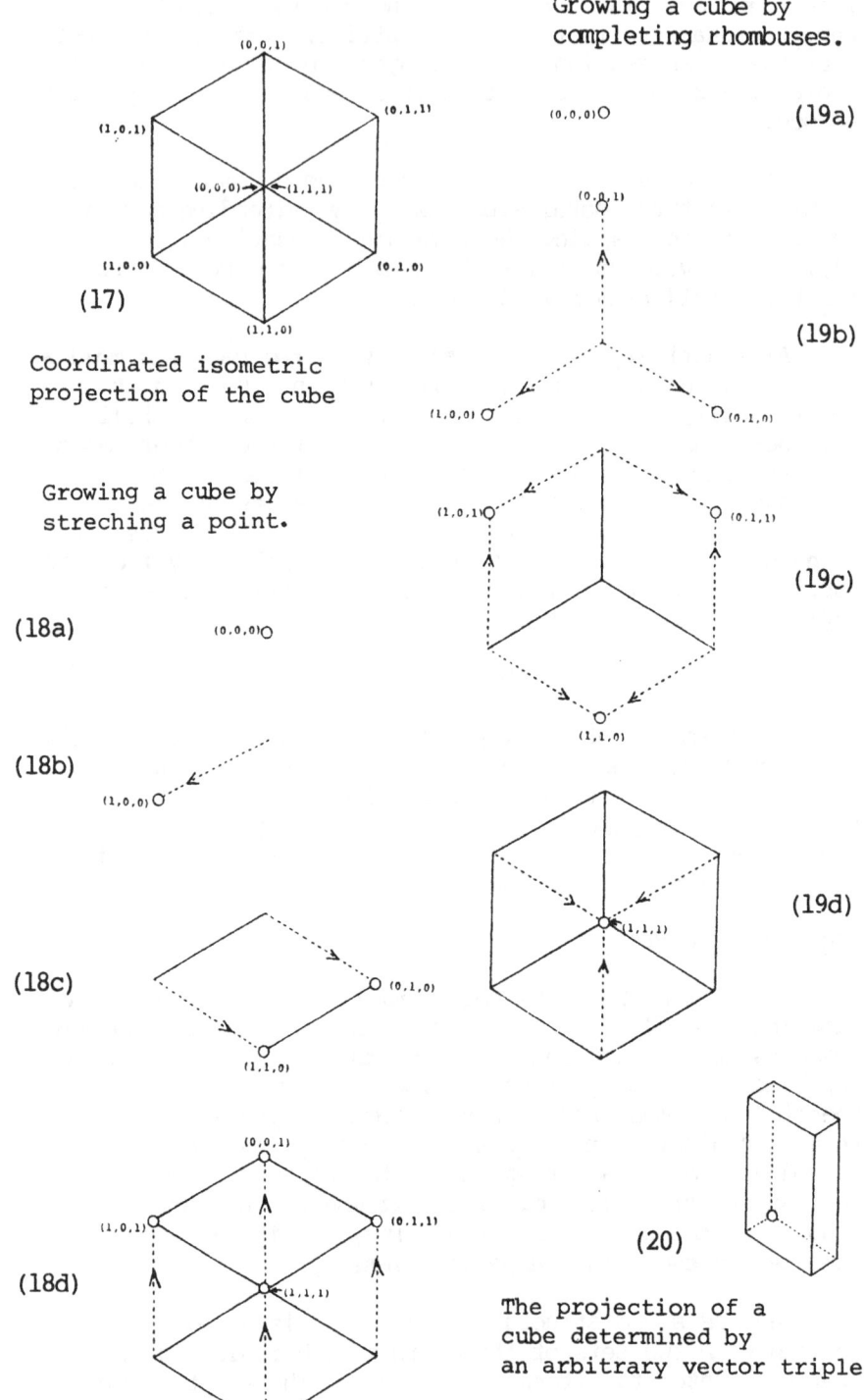

(17)

Coordinated isometric
projection of the cube

Growing a cube by
streching a point.

(18a)

(18b)

(18c)

(18d)

Growing a cube by
completing rhombuses.

(19a)

(19b)

(19c)

(19d)

(20)

The projection of a
cube determined by
an arbitrary vector triple.

vertices coincide at the center. This degeneracy at the center creates triangular loops of edges which are not present in the cube.

Six lines converge at the center of the hexagon. Three belong to the frontmost corner of the cube, the other three belong to the backmost corner. Note that the dotted edges Pa, Pb, Pc would normally be hidden by the front faces — they are visible only in a transparent cube. Together, the visible and invisible edges form three lines which connect opposite vertices of the hexagon.

The six square faces of the cube project onto six rhombuses with 120 degree obtuse angles. Three of the faces form the front (visible) surface (figure 14b), while the remaining three faces form the back (invisible) surface (figure 14c). Superimposing the two decompositions, we get a dissection of the hexagon into six triangles. Considered 2-dimensionally, the six rhombuses in figure 14a form the upper and lower layers of an uninflated balloon, joined only at the hexagonal perimeter. Notice how the rhombuses redistribute their edges along the perimeter. To draw a solid cube, an artist must decide which of the two layers hides the other.

The fact that figures 14b and 14c are rotations of each other forms the basis of an optical illusion. Surround a cube with three cubes each sharing one face with the central cube. In figure 15, the point P can either be seen "in", as the frontmost vertex of a central cube, or "out", as the meeting of the three cubes whose frontmost vertices are points a,b,c. In general, any crinkled surface (with no overhangs) has two such interpretations. The reversible staircase (figure 2b) is just one example (see also Escher's "Convex and Concave" in [Ernst]).

Figure 15 can be continued to produce an ambiguous tiling of the plane with rhombuses. The tiling can also be produced by projecting a whole plane of cubes as in figure 16. Here the fact that the hexagon is the shadow of the cube explains why it tiles the plane. Finally, if we let both the back and front faces show through, we obtain a triangular grid consisting of three sets of parallel lines meeting at three-way corners — the ideal matrix on which to draw cubical constructions.

Description by Coordinates

The unit cube in 3-space consists of all points whose coordinates lie between 0 and 1. Its vertices represent

all possible combinations of three 0's and 1's (figure
17). Since there are 3 coordinates, each with 2 possible
values, the total number of vertices must be 2x2x2=8. This
is consistent with the fact that the number of vertices in
a cube doubles with each new dimension.

Note that two vertices which differ in only one
coordinate are at opposite ends of the same edge. In
general, holding all but k number of coordinates fixed
produces a k-d face. Thus the faces of an n-cube are cubes
of lower dimensions.

If we view the unit cube down the line x=y=z toward
the origin, then the three unit segments (0,0,1), (0,1,0),
(1,0,0) project onto the segments Pa, Pb, Pc in figure 17.
Since our line of sight forms equal angles with the three
coordinate axes, the points a, b, and c are arranged
symmetrically in an equilateral triangle. Since Pa, Pb,
and Pc are congruent and at equal angles, the projection is
isometric.

How to Grow a Cube

The vertices of the isometric projection of the cube
can be evolved in two fundamentally different orders. Both
are important. The two procedures are best understood in
relation to coordinates.

Inductive Procedure In figure 18, a point (0-cube) is
successively stretched along various directions, producing
a line (1-cube), a square (2-cube), and finally a cube
(3-cube). At each step, the number of new vertices
doubles. The exact order in which the directions are
introduced is arbitrary. An n-dimensional cube can be
described as two parallel (n-1)-dimensional cubes which
have corresponding vertices connected by edges. An n-cube,
then, is a prism capped at either end by an (n-1)-cube.

A cube can be collapsed into a square along any one of
the three coordinate axes. From this fact, we can see that
a cube has 6 faces — two perpendicular to each axis. The
faces of the cube can also be counted by noting that a face
of the cube is isolated by holding one coordinate fixed at
either 0 or 1, and allowing all other coordinates to vary
through all combinations.

The inductive approach emphasizes the rectilinear
nature of the cube — the dimensions are clearly seen
entering one by one. This asymmetry makes it easy to grasp
the parts of the cube, but works against the particular

symmetry of the isometric projection.

Combinatorial procedure   Figure 19 starts with three
unit line segments radiating symmetrically from a point and
proceeds to combine them in pairs to form rhombuses. At
each step, whenever two unit line segments meet at a
vertex, the remaining two edges of the rhombus are drawn in.

In this procedure the cube grows symmetrically from a
corner, rather than doubling exponentially. Note that
faces are introduced in order of increasing
dimensionality. Initially there is only a vertex
(0-face). Then the adjacent edges (1-faces) are
introduced. Edges are grouped in pairs to determine
rhombuses (2-faces). Finally, the 3 front faces not
adjacent to the original vertex are filled in, completing
the cube (3-face). Vertices are introduced in order of
increasing distance from the origin, with (1,1,1) entering
last. Note that figure 19a lists all vertices with exactly
0 coordinates at value 1, figure 19b lists all vertices
with exactly 1 coordinate at value 1, figure 19c lists all
vertices with exactly 2 coordinates at value 1, and so on.

The combinatorial approach emphasizes the symmetrical
nature of the cube — all dimensions are equally present
from the beginning. This interpretation accurately
reflects the symmetry of the isometric projection, but
makes it difficult to see the components.

We have described two methods for deriving the
isometric projection of the 3-d cube into a 2-plane from
three translation vectors. These methods will work equally
well for non-isometric projections. In general, any three
vectors can be used to generate a picture of a cube, as
illustrated in figure 20. Since the projection is
orthogonal, edges parallel in the original cube map onto
parallel line segments. Note that the projections of
(0,0,0) and (1,1,1) do not coincide. Additionally,
projections need not be restricted to 2 dimensions. If all
three translation vectors are colinear, then we obtain the
twice-collapsed picture of a cube as seen by a 2-d person.
If the three translation vectors are at mutual right
angles, then we obtain the ultimate picture of a cube —
the 3-dimensional object itself.

Vision

Earlier we began listing the assumptions which
underlie vision. In this section, we will continue
unraveling the elements of vision, to understand how the

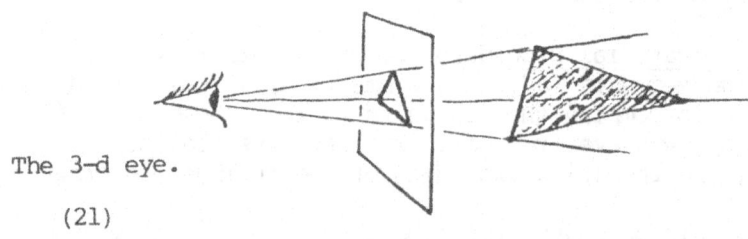

The 3-d eye.

(21)

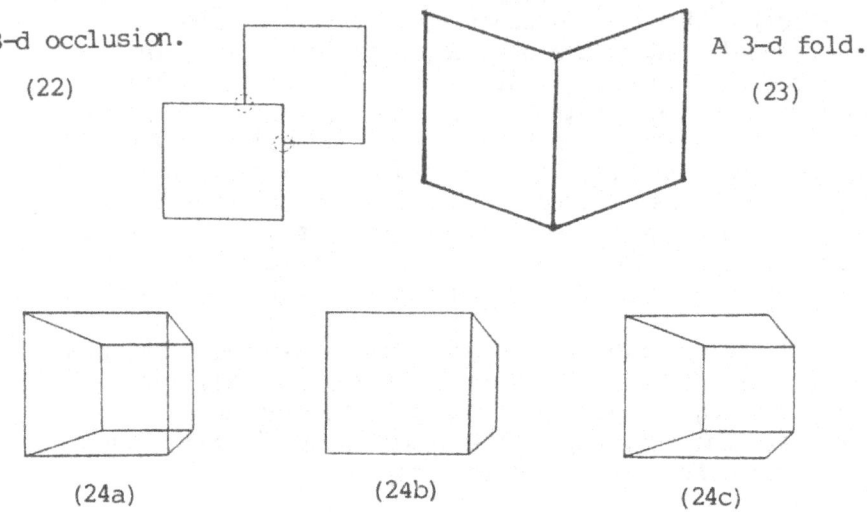

3-d occlusion.

(22)

A 3-d fold.

(23)

(24a)   (24b)   (24c)

A wire-frame drawing;
two possible interpretations.

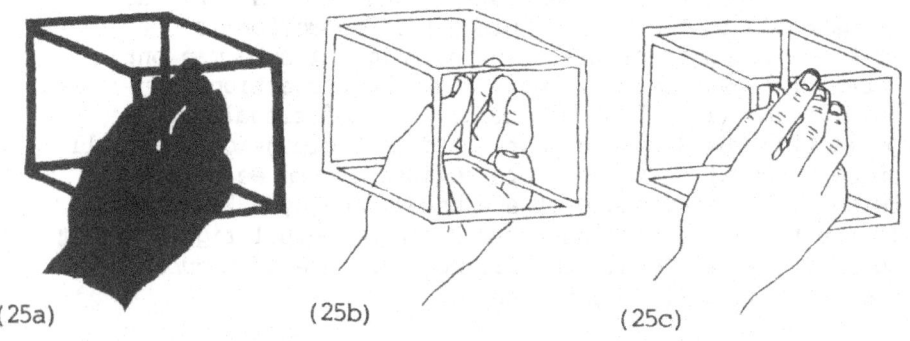

(25a)   (25b)   (25c)

The shadow of a hand
holding a Necker cube;
two possible interpretations.

eye is able to pull 3 dimensions of information out of a
2-d picture. The study is by no means complete, but will
be sufficient to allow us to jump to 4-space.

3-d Vision

In classical perspective, the canvas is a metaphorical
window intersecting the pyramid of visual rays whose apex
is the eye of the observer (the "viewpoint"), and whose
base is the object being represented (figure 21). Since
the procedure is an exact tracing, the drawing and the
object create the same 2-d image on the back of the eye.
In theory, a viewer standing in the same relation to the
canvas as the original artist should see a copy of the
original scene.

Classical perspective has several important features.
Straight lines always map to straight lines (except for
lines which point directly at the viewpoint, which map to
points). Angles are not usually preserved. Note how
folding the two squares shown in figure 22 has distorted
them into parallelograms. Figure 23 portrays two squares,
one partially blocking the other. We in 3 dimensions
understand that the hidden portion of the rear square
continues behind the front square. A 2-d person would not
understand where there is room to "go underneath", and
would interpret the rear square as having a bite taken out
of it.

Convex shells always deflate to surfaces of double
thickness. For instance, of the 6 quadrilaterals in figure
24a representing the faces of the cube, 2 form the front
surface, and the remaining 4 form the back. The two
surfaces are joined at the perimeter, allowing the
"balloon" to be reinflated without losing air. If the
shell is opaque, then the set of front faces will
completely hide the set of back faces.

Perspective introduces size distortion, which varies
with the distance between the object and viewpoint. Since
the front face of the cube portrayed in figure 24a appears
to be much larger than the back face, we can say that the
viewpoint is quite close to the cube. The back face is
also shifted to the right with respect to the front face,
so we can add that the viewpoint must have been chosen
slightly off-center. As the viewpoint moves farther away,
the perspective distortion will decrease until it is
effectively zero. This is why images appear flat when
viewed through a telescope.

For our purposes, we will not be interested in
perspective distortion, so the viewpoint will be put at
infinity. Such a perspectiveless projection is called
"orthographic". In orthographic projections, parallel
lines map to parallel lines. There are no vanishing
points. Objects remain the same size regardless of
distance.

Orthography is a good simplifying assumption, but can
create problems of its own. Figure 24a can be interpreted
either as a cube (figure 24b), or, removing the other set
of hidden lines, as a truncated square pyramid pointing
towards the viewer (figure 24c). Since cubes are as a rule
more common (and more symmetric) than truncated pyramids,
the cubical interpretation is preferred. In the
orthographic projection shown in figure 25a (the Necker
cube), however, front and back faces are the same size.
Neither interpretation is given priority; both are equally
likely (figures 25b and 25c). This ambiguity is a
fundamental property of all orthographically projected
shadows. For example, because the sun is in effect
infinitely far away, you may think of your shadow as either
looking towards or away from you.

How is the eye able to see a 3-d object in a 2-d
picture? We have already mentioned the cues of perspective
size distortion and hidden line removal. Other depth cues
include "T-intersections" (circled in figure 23), focus,
atmospheric diffusion, stereoscopy, parallax, and feedback
from other senses. Ultimately, though, the missing
dimension of depth can never be completely restored. Any
point P in a drawing might stand for any of the infinitely
many points which lie on the line connecting P and the
viewpoint. We are able to judge the size of people in
photographs only because people and other visual stimuli
tend to occur in certain predictable varieties.

Vision, then, is a learned skill. Translating 2-d
images into 3-d objects requires the aid of an acquired
vocabulary of expected patterns. Different vocabularies
produce different translations. Ambiguity is unavoidable,
due to the information loss in projecting 3 dimensions into
2. We have mentioned only some of the highest level
skills. There are many more levels. Researchers in
artifical intelligence have had to characterize the
expectations needed in order to just recognize the lines
and angles which make up a line drawing [Winston].

One last question remains before we go on to 4
dimensions. What does it mean to find the 4-dimensional

analog of a 3-dimensional object? There might be many generalizations; how do we decide on just one?

Analogy

Analogy is like translation. In order to say a sentence in a different language, the translator must determine which features are to be preserved (content), and which are merely artifacts of the language (context). Different languages produce different translations. If the languages are similar, a word-for-word translation may be possible. If the languages are different, it may be necessary to rethink the sentence to find how its expression depends on the particular language. Translation requires a careful understanding of the relation between language and meaning. Conversely, any translation reveals how the translator understands language. The main difficulty of translation is recognizing that there are in fact two separate levels. For as long as we stay within one language, there is no need to distinguish between context and content.

The main difficulty in generalizing an object to 4 dimensions is recognizing dimensionality itself — the absurd concept of assigning a number to space. 3-space is as real and unreal as is 4-space. Once dimensionality has been isolated as purely formal property, it is easy to change its value.

The context for understanding the impossible triangle is 3-d visual perception. The content is a certain play with depth perception based on the properties of the cube. To translate the impossible triangle into 4-d space, we must first understand which properties of vision are dependent on dimensionality. Once our knowledge has been regrouped, we can "add 1" to the impossible triangle. This regrouping may seem strange, since we do not normally recognize dimensionality as a property which can be easily "slipped". But we do have experience with 2-d space. Meditating on the relation between 2 and 3 dimensions gives us a running start towards 4. If it were not for that relation, understanding 4-space would be hopeless. We are indeed fortunate to live in a space of multiple spatial dimensions.

4-d Vision

In classical 4-d perspective, the canvas is a metaphorical window intersecting the pyramid of visual rays whose apex is the eye of the observer (the "viewpoint") and

whose base is the object being represented (figure 26).
Since the procedure is an exact tracing, the drawing and
the object create the same image on the back of the eye.
In theory, a viewer standing in the same relation to the
canvas as the original artist should see a copy of the
original scene.

Classical perspective has several important features.
Straight lines map to straight lines, flat planes map to
flat planes (except for lines and planes which point
directly at the viewpoint, which map to points and lines,
respectively). Angles are not usually preserved. Note how
folding the two squares shown in figure 27 has distorted
them into parallelepipeds. Figure 28 portrays two cubes,
one partially blocking the other. We in 4 dimensions
understand that the hidden portion of the rear cube
continues behind the front cube. A 3-d person would not
understand where there is room to "go underneath", and
would interpret the rear cube as having a bite taken out of
it.

Convex shells always deflate to 3-surfaces of double
thickness. For instance, of the 8 hexahedra in figure 29a
representing the 3-faces of the 4-cube, 3 form the front
surface, and the remaining 5 form the back. The two
surfaces are joined at the 3-perimeter, allowing the
"balloon" to be reinflated without losing air. If the
shell is opaque, then the set of front 3-faces will
completely hide the set of back 3-faces.

Perspective introduces size distortion, which varies
with the distance between the object and viewpoint. Since
the front 3-face of the 4-cube portrayed in figure 29a
appears to be much larger than the back 3-face, we can say
that the viewpoint is quite close to the 4-cube. The back
3-face is also shifted to the right with respect to the
front 3-face, so we can add that the viewpoint must have
been chosen slightly off-center. As the viewpoint moves
farther away, the perspective distortion will decrease
until it is effectively zero. This is why images appear
flat when viewed through a telescope.

For our purposes, we will not be interested in
perspective distortion, so the viewpoint will be put at
infinity. Such a perspectiveless projection is called an
"orthographic" projection. In orthographic projections,
parallel lines and planes map to parallel lines and
planes. There are no vanishing points. Objects remain the
same size regardless of distance.

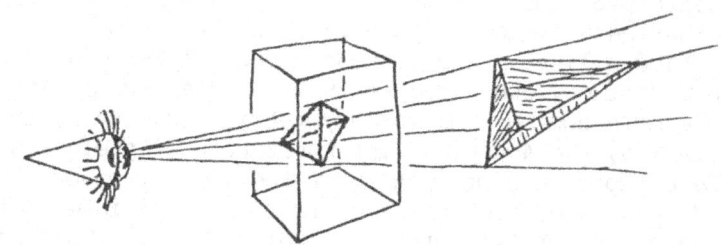

The 4-d eye.

(26)

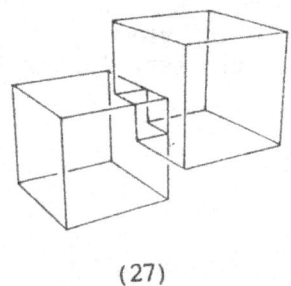

(27)

4-d occlusion.

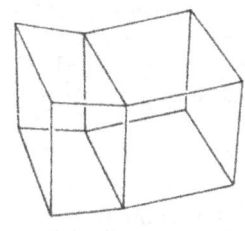

(28)

A 4-d fold.

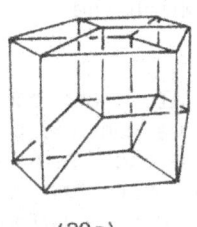

(29a)

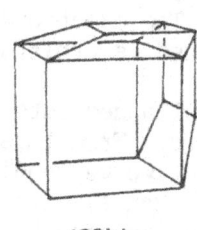

(29b)

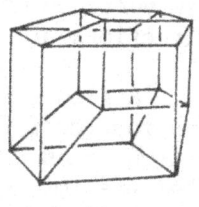

(29c)

A wire-frame drawing;
two possible interpretations.

Orthography is a good simplifying assumption, but can create problems of its own. Figure 29a can be interpreted either as a 4-cube (figure 29b), or, removing the other set of hidden lines, as a truncated cubical pyramid pointing towards the viewer (figure 29c). Since 4-cubes are as a rule more common (and more symmetric) than truncated pyramids, the 4-cubical interpretation is preferred. In the orthographic projection of the 4-cube, however, front and back faces are the same size. Neither interpretation is given priority; both are equally likely. This ambiguity is a fundamental property of all orthographically projected shadows. For example, because the 4-sun is in effect infinitely far away, you may think of your shadow as either looking towards or away from you.

How is the eye able to see a 4-d object in a 3-d picture? As in 3-d perspective, the missing dimension of depth can never be completely restored. Any point P in a drawing might stand for any of the infinitely many points which lie on the line connecting P and the viewpoint. 4-d vision, like 3-d vision, is a learned process, dependent on a vocabulary of expected patterns.

The 4-d Cube

How to Grow a 4-cube

The vertices of the isometric projection of the 4-cube can be evolved in two fundamentally different orders. Both are important. The two procedures are best understood in relation to coordinates.

Inductive Procedure   In figure 30, a point (0-cube) is successively stretched along various directions, producing a line (1-cube), a square (2-cube), a cube (3-cube), and finally a hypercube (4-cube). At each step, the number of new vertices doubles. The exact order in which the directions are introduced is arbitrary.

A 4-cube can be collapsed into a cube along any one of the four coordinate axes. From this fact, we can see that a 4-cube has 8 faces — two perpendicular to each axis. The faces of the 4-cube can also be counted by noting that a face of the 4-cube is isolated by holding one coordinate fixed at either 0 or 1, and allowing all other coordinates to vary through all combinations.

The inductive approach emphasizes the rectilinear nature of the 4-cube -- the dimensions are clearly seen entering one by one. This interpretation gives the

Growing a 4-cube by
stretching a point.

Growing a 4-cube by
completing rhombuses.

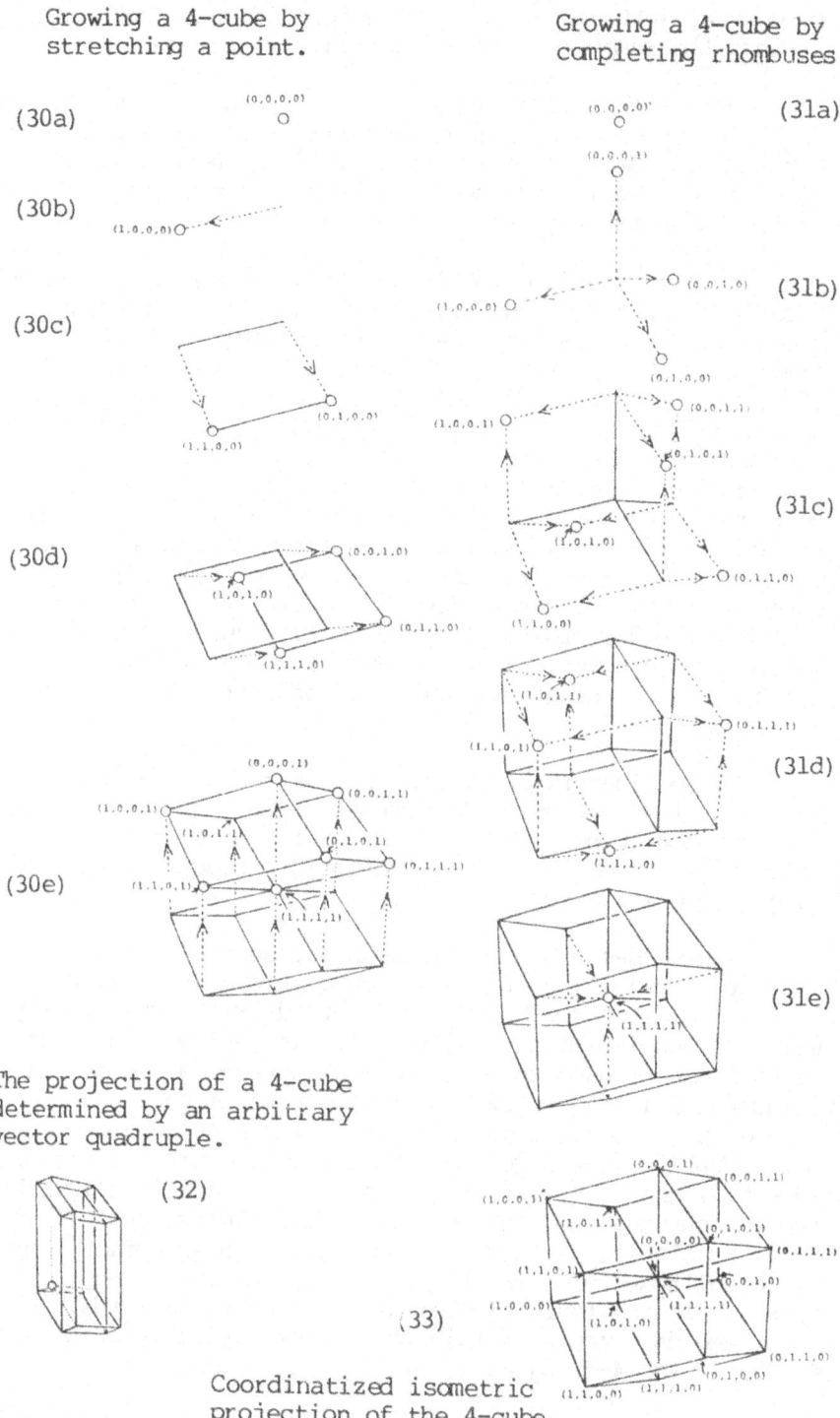

The projection of a 4-cube
determined by an arbitrary
vector quadruple.

Coordinatized isometric
projection of the 4-cube.

perceptual intuition an asymmetry to hold onto, but works against the symmetry of the isometric projection.

<u>Combinatorial Procedure</u> Figure 31 starts with four unit line segments radiating symmetrically from a point and proceeds to combine them in pairs to form rhombuses. At each step, whenever two unit line segments meet at a vertex, the remaining two edges of the rhombus are drawn in.

In this procedure the 4-cube grows symmetrically from a corner, rather than doubling exponentially. Note that faces are introduced in order of increasing dimensionality. Initially there is only a vertex (0-face). Then the adjacent edges (1-faces) are introduced. Edges are grouped in pairs to determine rhombuses (2-faces), and in triples to determine rhombohedra (3-faces). Finally, the 4 front 3-faces not adjacent to the original vertex are filled in, completing the 4-cube (4-face). Vertices are introduced in order of increasing distance from the origin, with (1,1,1,1) entering last. Note that figure 31a lists all vertices with exactly 0 coordinates at value 1, figure 31b lists all vertices with exactly 1 coordinate at value 1, figure 31c lists all vertices with exactly 2 coordinates at value 1, and so on.

The combinatorial approach emphasizes the symmetrical nature of the 4-cube — all dimensions are equally present from the beginning. This interpretation accurately reflects the symmetry of the isometric projection, but is more difficult to see.

We have described two methods for deriving the isometric projection of the 4-d cube into a 3-plane from four translation vectors. These methods will work equally well for non-isometric projections. In general, any four vectors can be used to generate a picture of a 4-cube, as illustrated in figure 32. Since the projection is orthogonal, all edges parallel in the original 4-cube map onto parallel line segments. Note that the projections of (0,0,0,0) and (1,1,1,1) do not coincide. Additionally, the projections need not be restricted to 3 dimensions. If all four translation vectors are coplanar, then we obtain the twice-collapsed picture of a 4-cube as seen by a 3-d person. If the four translation vectors are at mutual right angles, then we obtain the ultimate picture of a 4-cube — the 4-d object itself.

Description by coordinates

The unit cube in 4-space consists of all points whose coordinates lie between 0 and 1. Its vertices represent all possible combinations of four 0's and 1's (figure 33). Since there are four coordinates, each with 2 possible values, the total number of vertices must be 2x2x2x2=16. This is consistent with the fact that the number of vertices in a cube doubles with each new dimension.

Note that two vertices which differ in only one coordinate are at opposite ends of the same edge. In general, holding all but k number of coordinates fixed produces a k-dimensional face. Thus the faces of an n-cube are cubes of lower dimensions.

If we view the unit cube down the line w=x=y=z toward the origin, then the four unit segments (0,0,0,1), (0,0,1,0), (0,1,0,0), (1,0,0,0) project onto the segments Pa, Pb, Pc, Pd in figure 33. Since our line of sight forms equal angles with the four coordinate axes, the points a, b, c, d are arranged symmetrically in an equilateral triangle. Since Pa, Pb, Pc, Pd are congruent and at equal angles, the projection is isometric.

The isometric projection

Figure 34a shows the most symmetric projection of the 4-cube into 3-space. All edges in this projection map onto segments of the same length. Thus the projection is isometric — the four coordinate axes of space are evenly distributed around a sphere, radiating out to the corners of a regular tetrahedron abcd. This extreme 3-d uniformity simplifies the processing of drawing, but it can also confuse the eye with unexpected symmetry.

Seen as a 3-d structure, figure 34a is a rhombic dodecahedron with two sets of four lines each radiating out from the center to alternate vertices. "Dodeca" for twelve faces, and "rhombic" to distinguish it from the more common dodecahedron with pentagonal faces. Unlike the hexagon, the rhombic dodecahedron is not considered a regular polytope. Nonetheless, all its faces are equivalent.

The sixteen vertices of the original 4-cube project in two different ways. Fourteen of the vertices form the

rhombic dodecahedron, while the remaining two vertices
coincide at the center. This degeneracy at the center
creates quadrangular loops of edges which are not present
in the 4-cube.

Eight lines converge at the center of the rhombic
dodecahedron. Four belong to the frontmost corner of the
4-cube, the other four belong to the backmost corner. Note
that the four dotted edges Pa, Pb, Pc, Pd would normally be
hidden by the front 3-faces — they are visible only in a
transparent 4-cube. Together, the visible and invisible
edges form four lines connecting opposite corners of the
rhombic dodecahedron.

The eight cubical 3-faces of the 4-cube project onto
eight rhombohedra with 120 degree dihedral angles. Four of
the 3-faces form the front (visible) surface (figure 34b),
while the remaining four 3-faces form the back (invisible)
surface (figure 34c). Superimposing the two
decompositions, we get a dissection of the rhombic
dodecahedron into 24 (non-regular) tetrahedra. Considered
3-dimensionally, the eight rhombohedra in figure 34a form
the upper and lower layers of an uninflated balloon, joined
only at along the rhombic dodecahedral perimeter. Notice
how the rhombohedra redistribute their 2-faces along the
perimeter. To draw a solid 4-cube, an artist must decide
which of the two layers hides the other.

The fact that figures 34b and 34c are rotations of
each other forms the basis of an optical illusion.
Surround a 4-cube with four 4-cubes each sharing one 3-face
with the central 4-cube. In figure 35, the point P can
either be seen "in", as the frontmost vertex of a central
4-cube, or "out", as the meeting of the four 4-cubes whose
frontmost vertices are points a,b,c. In general, any
crinkled surface (with no overhangs) has two such
interpretations. The 4-d analog of the reversible
staircase (figure 2b) is just one example.

Figure 35 can be continued to produce an ambiguous
packing of space with rhombohedra. The packing can also be
produced by projecting a whole 3-plane of 4-cubes (figure
36). Here the fact that the rhombic dodecahedron is the
shadow of the 4-cube explains why it packs space. Finally,
if we let both the back and front 3-faces show through, we
obtain a skew quadrangular grid consisting of four sets of
parallel lines meeting at four-way corners — the ideal
matrix on which to draw 4-cubical constructions.

Additional information on the rhombic dodecahedron and

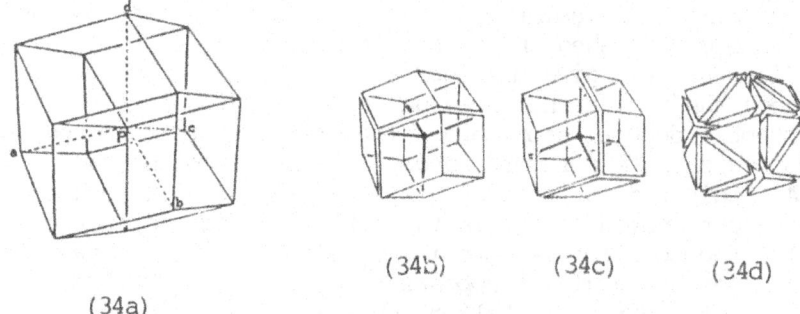

(34a)

(34b)   (34c)   (34d)

The isometric projection
of the 4-cube;
front and back faces.

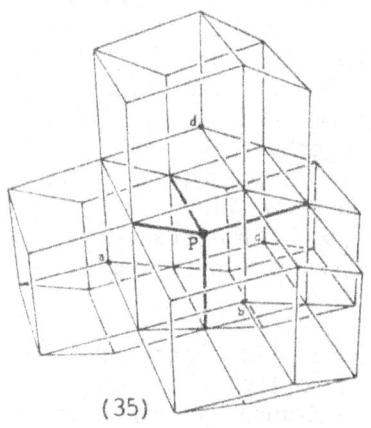

(35)

Excerpt from an
ambiguous rhombohedral
packing.

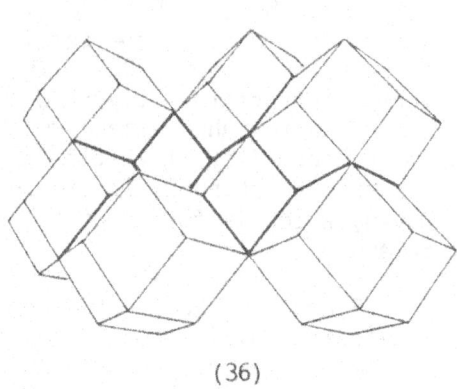

(36)

Packing space with
rhombic dodecahedra.

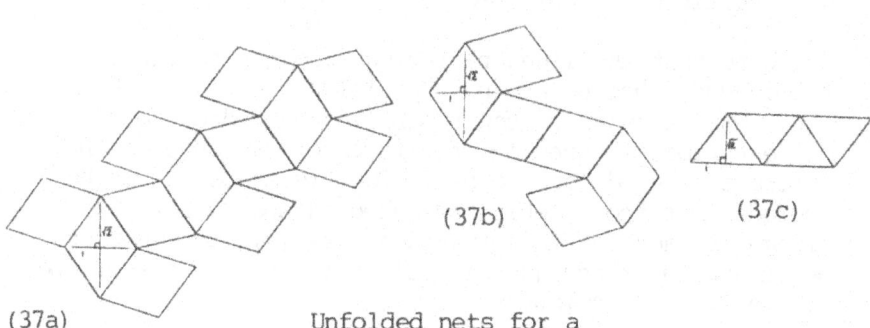

(37a)

(37b)   (37c)

Unfolded nets for a
rhombic dodecahedron,
rhombohedron, and
tetrahedron.

other polyhedra can be found in [Coxeter], [Critchlow], [Hilbert] and [Holden]. The best way to understand the rhombic dodecahedron is to live with a model of it. Figures 37abc give unfolded nets for the rhombic dodecahedron, a rhombohedron, four of which make the rhombic dodecahedron, and a (non-regular) tetrahedron, six of which make a rhombohedron. All folds go the same way. As indicated in the diagrams, the diagonals of each rhombus are perpendicular and in the ratio 1 to square root of 2. The rhombohedron is especially deceptive since viewed from the correct angle it appears to be a cube. The faces of the tetrahedron are half-rhombuses. The reader is encouraged to play with these models to get a feel for the different ways a rhombic dodecahedron can be decomposed.

Another way to understand the rhombic dodecahedron is in relation to a cubical grid. The tetrahedron has four vertices; the cube has eight. If we leave out every other vertex of the cube, the remaining four vertices form a tetrahedron (figure 38a). The four omitted vertices also form a tetrahedron (figure 38b). Together, the two inscribed tetrahedra make up the "stella octangula", which can be formed by stellating an octahedron (figure 38c). The stella octangula is the 3-d analog of the Star of David, which is formed of two complementary overlapping triangles.

Since the rhombic dodecahedron is also generated from a tetrahedron, it too can be related to the cube. In figure 39, the tetrahedron abcd from figure 38a has been rotated to match the vertices of a cube. Within the rhombic dodecahedron, the 6 white vertices form an octahedron, while the 8 black vertices form a cube. Note that the white vertices are 4-way junctions, while the black vertices are only 3-way.

Note that the lines radiating from the center of the rhombic dodecahedron are now spatial diagonals of the cube. The 24 edges of the rhombic dodecahedron form six square pyramids connecting the faces of the cube to the centers of the adjacent cubes. On this basis one might conclude that the rhombic dodecahedron should have 24 triangular faces — four for each pyramid. But in fact, the triangular faces join in pairs at the 12 edges of the cube to form 12 rhombuses.

The fact that the tetrahedron can be related so nicely to the cube is an important idiosyncrasy of 3-space, and should not be taken for granted. There is no equally symmetric way to align the 3 vertices of the triangle

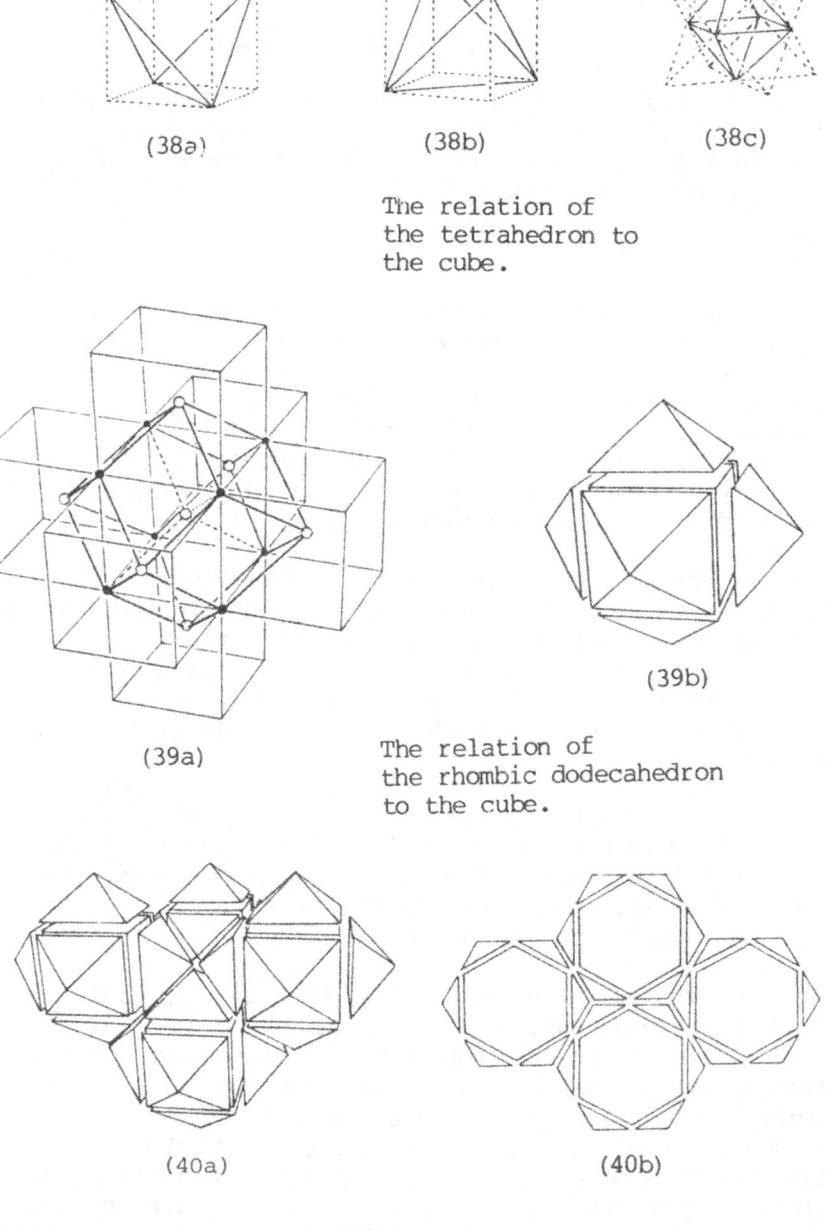

(38ə)   (38b)   (38c)

The relation of
the tetrahedron to
the cube.

(39b)

(39a)

The relation of
the rhombic dodecahedron
to the cube.

(40a)   (40b)

Another way to think
of the rhombic dodecahedral space packing;
the 2-d analog.

(2-tetrahedron) with the square (2-cube). The triangle is related to the cube in 3-space only because the number of vertices happens to be a power of 2.

Elements of Construction

The elements used in constructing the 4-d analog of the impossible triangle follow their 3-dimensional counterparts closely. Once the model has been completed, we will offer an impossibility proof.

Beams

A cubical beam is formed by taking a cube and translating it along a perpendicular axis. This sounds very much like the definition of a 4-cube, and indeed, a 4-cube is a very short cubical beam. Note that the "cubical" in "cubical beam" gives the shape of its cross-section, just as in the term "cubical prism". The word "beam" is used instead of "prism" only because it has fewer dimensional connotations.

Figure 41 shows a picture of a cubical beam with hidden lines removed. The horizontal plane cutting through the center emphasizes that this is a strictly 3-d figure. Note that the cross-section of the projection of the cubical beam looks like a hexagon with 3 lines radiating to alternate vertices — a picture of a cube as seen by a 3-d person.

The four visible faces of the beam decompose the long column into four parallelepipeds — three extending down its length, and one capping the remaining concavity (figure 42). If we were to shrink the length of the beam to zero, all that would remain would be the cap.

Alternatively, we could start with the cap, and mold a copy of the undersurface (three rhombuses meeting at obtuse angles, i.e. half a rhombohedron). This concave shell will nest smoothly under the cap. Take nine very wide rubber bands and have them join each of the nine edges of the undersurface of the cap to the nine edges of the shell. Then an image of a beam (as well as an inadvertent sling shot) is generated by pulling the shell away from the cap. Note that this construction is strictly 3-d — no explicit reference is made to 4-space.

Corners

The most important component of the 4-d analog of the

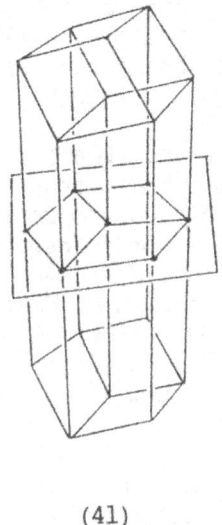

(41)

Sliced cubic beam.

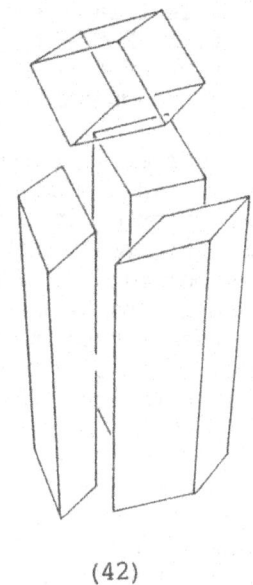

(42)

Exploded cubic beam.

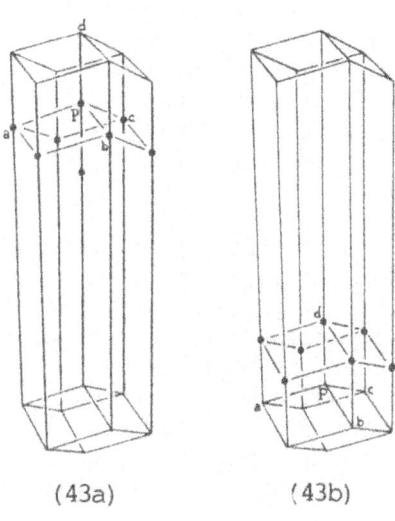

(43a)   (43b)

The two ways of extending
a 4-cube into a beam.

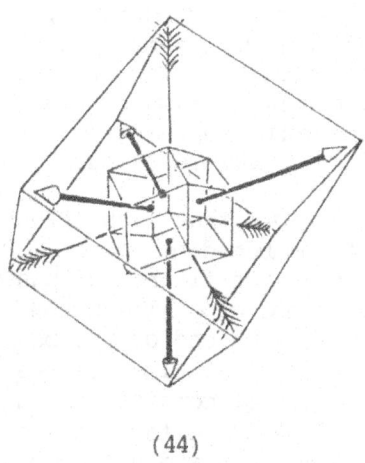

(44)

The eight directions for
extending a 4-cube into
a beam.

impossible triangle will be the corner. The way corners
twist over themselves is responsible for 4-dimensionality,
the effect of continual recession. They are also hard to
draw. Consequently, the corner needs careful study.

A corner is created whenever two cubical beams
intersect in a 4-cube. Alternately, we can construct a
corner by extending a 4-cube in two directions to create
beams. Figure 45abcd shows the four types of corners which
can be pulled from the 4-cube pictured in figure 45e. The
black dots and broken lines outline vertices and edges of
the corner 4-cube which have been removed by the extended
beams. It is useful to superimpose figure 45e on each of
the illustrations to get a getter idea of how edges enter
and leave the corner.

Note that corner 45b can be considered the "back" of
corner 45a. Similarly, corner 45d reverses the relation of
hidden and unhidden lines from corner 45c. We can even
consider that a surface which folds around the back in one
corner is continued in the sibling corner. Corners 45b and
45d appear less natural than other two, since they are seen
from underneath.

As 4-d structures, the four corners can be explained
as four different views of a single right angle. As 3-d
pictures, however, they differ in which faces are visible.
The distribution of visible faces, in turn, determines how
the corner appears to twist. Let us study the four types
of 4-d corners as purely 3-d structures.

Note that each illustration has a different assortment
of rhombohedral "caps". Figure 45c has a cap at either
end, while figure 45d has none. Both figures 45a and 45b
have one cap. The presence of a cap means that we are able
to see the end of the beam, i.e. that the beam is
approaching. The absence of a cap indicates a receding
beam. All combinations are represented here. Figures 45c
and 45d are either "all-positive" or "all negative", while
figures 45a and 45b combine both approaching and receding
beams.

We may also study the way the visible surfaces are
redistributed as they pass around a corner. Coming into a
corner there are six "surface bands", corresponding to the
six faces of the cubical cross section. Only three bands
are visible at a time, the opposite three bands being
hidden on the back. At a corner, there are two
possibilities. Either the bands will retain their
ordering, or they will undergo a spherical permutation in

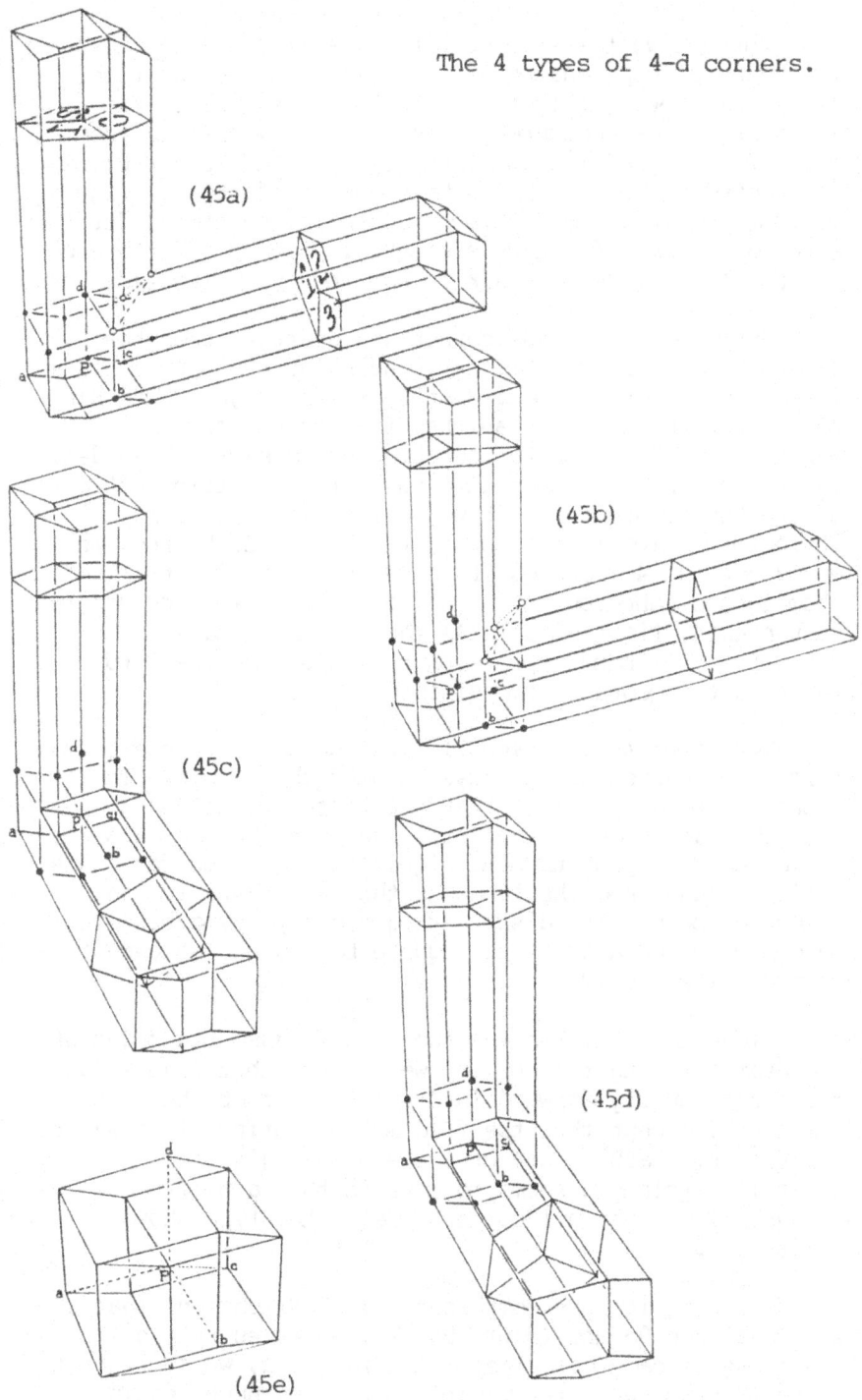

The 4 types of 4-d corners.

(45a)

(45b)

(45c)

(45d)

(45e)

which one band vanishes and a new one appears.

Suppose we traverse corner 45c starting from the lower
rhombohedral cap, ending at the upper. As we enter the
corner, we bend gently to the right. The band on the
innermost track encounters a fold plane (emanating from
point P), but retains its innermost position after being
"refracted". The outermost bands angle off to the right
without folding. The surfaces leave in the same order they
entered. Corner 45d acts similarly. The only difference
is that the outermost track is now the one refracted.

Suppose we traverse corner 45a starting from the
(invisible) right rhombic cap, ending at the upper. As we
enter the corner, we bend sharply to the right. The band
on the innermost track vanishes from sight abruptly at the
corner. The bands on the outermost tracks switch suddenly
to the innermost as they turn the corner, cutting off a tip
of the innermost band. If we were to add all such
tetrahedral tips to the corner rhombic dodecahedron, we
would get the 4-d analog of the Star of David (stellated
rhombic dodecahedron; shown on top of the right column in
"Waterfall", figure 55a). As the innermost band is being
cut off, a new band appears on the outermost track to
maintain the quota of two visible bands.

Described 4-dimensionally, we can say that the beam
twists 90 degrees as it passes around the corner. The
appearing band signifies that a surface has folded around
from the back side to the front where it can now be seen,
while the disappearing band signifies that a surfaces has
folded around from the front to the back where it can no
longer be seen. The disappearing and appearing bands occur
on opposite surfaces of the square beams, and so cannot be
seen simultaneously.

Since the beam leaving the corner passes in front of
the beam entering the corner, we can say that figure 45a
moves continually forward in 4-space. Corner 45b acts
similarly, except that the fold moves continually away from
the viewer. Each corner of the 4-d analog of the
impossible triangle seems to move farther away from the
viewer. Therefore all corners are of the type shown in
figure 45b.

Twisting is more complicated in 3 dimensions than in
2. In figure 9c and figure 9d, all surfaces retain their
order, as already described. In figure 9a, we see that the
entering beam has three visible surfaces, numbered in
increasing order clockwise (modulo 6). This property is

not retained in the exiting beam. Instead, the numbers
decrease when read clockwise. Thus successive corners will
have alternate handedness.

Global form

What will be global shape of the 4-d analog of the
impossible triangle? One way to derive the shape is to
rotate a 4-cube until two opposite vertices appear to
coincide. Figure 46 shows such a 4-cube, with four edges
darkened. Imagine the 4-d structure formed by this
sequence of edges. The point of view has been chosen so
that two opposite vertices of the 4-cube appear to coincide
in the center in this 3-d projection. As a result, the
sequence of darkened edges appears to form a closed loop,
when in fact the two ends are at different depths in
4-space. Furthermore, the angle of the spurious corner is
exactly like the other three legitimate corners — the loop
is an equiangular quadrilateral. A devious artist could
retouch this corner to imitate the other three, concealing
the false connection.

Another way to derive the shape is to assign
coordinates to the vertices. If the beginning vertex in
back is $(0,0,0,0)$ and the ending vertex in front is
$(1,1,1,1)$, then the point of view looks directly down the
line $w=x=y=z$ towards the origin. The paradox is that
having traveled a unit distance in turn along the w, x, y
and z directions you return to the origin, not $(1,1,1,1)$.

The global form of the 4-d analog of the impossible
triangle, then, is constructed by rearranging the unit
vectors in the w,x,y,z directions to form a closed loop.
We have chosen our projection so that the four basis
vectors radiate symmetrically out to the corners of a
regular tetrahedron (figure 47a). Rearranging these
vectors, we obtain a skew, but equiangular, equilateral
quadrangle (figure 47b).

Note that in this case the 3-d analog of the triangle
is a quadrilateral, not a tetrahedron. An impossible
tetrahedron may still be possible, but only if the shape of
the impossible triangle is derived in a totally different
way.

The Impossible Skew Quadrilateral

Models

Figure 48 shows the completed model of the impossible

skew quadrilateral. Remember that this is a 3-d picture of
an apparent 4-d object. Only the edges have been retained
in this 2-d projection. A 4-d person would not consider
the figure adequate until its planar surfaces were filled
in, just as a 3-d person would not consider the vertices of
the impossible triangle to be an adequate optical illusion
until its edges were filled in. A 4-d person might also
request that the various 3-d surfaces of the impossible
skew quadrilateral be assigned different colors to suggest
a light source somewhere off in 4-space. Thus a proper 3-d
model should consist of several solid blocks of color, not
just an empty frame.

Let us leave 4-space and treat figure 48 as a purely
3-d structure. Figure 49abcd shows how the impossible skew
quadrilateral can be built from four congruent pieces. The
pieces are marvelously deceptive for the 3-d eye. We
expect right angles and get multiples of 60 degrees
instead. In fact, individual pieces, viewed from the
correct angle, almost appear to be impossible triangles.

Each piece traces a large " $\angle\!\!\!\_\rfloor$ ", with the three arms
running along three of the four sides of the skew
quadrangle. The arrangement is symmetrical — mapping one
piece into another rotates the structure into itself (a
mirror reflection about a plane perpendicular to the axis
of symmetry may also be necessary). If you think of the
four corners of quadrangle as four horses on a carousel,
each of which makes two complete up/down motions during the
course of one carousel revolution, then the mapping is
equivalent to turning the carousel through one quarter of a
revolution (figure 50a).

A 3-d person wishing to understand the impossible skew
quadrangle might very well make such a 4-piece model, since
only a model which comes apart allows a view of what
happens on the hidden internal surfaces. Note that each
piece is composed of a series of rhombohedra corresponding
to cubical surfaces on the original cubical beams (figure
50b). The one irregular end is clipped when one beam
passes in front of another beam in 4-space.

The unfolded plan for an individual piece is given in
figure 51. Solid lines indicate convex folds, while the
one double line indicates a concave fold. Dotted lines are
to be cut. Thin regions serve as tabs. These get rather
bulky, especially on smaller models. It may be better to
leave them out. All apparent rhombuses are congruent and
have diagonals in a ratio of 1 to square root of 2.
Reverse all the folds to produce the mirror image of a

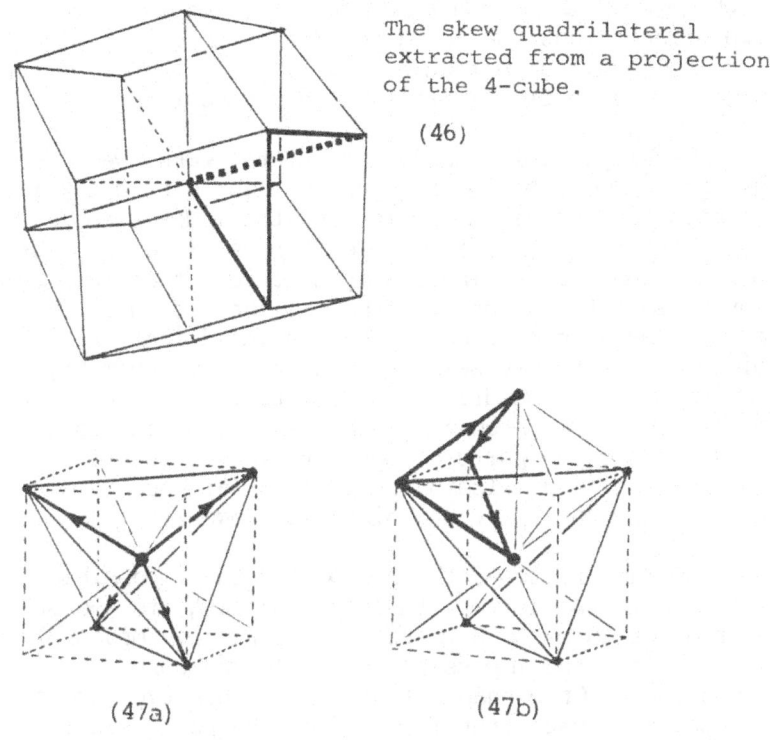

The skew quadrilateral
extracted from a projection
of the 4-cube.

(46)

(47a)   (47b)

Skew quadrilateral =
rearranged basis vectors.

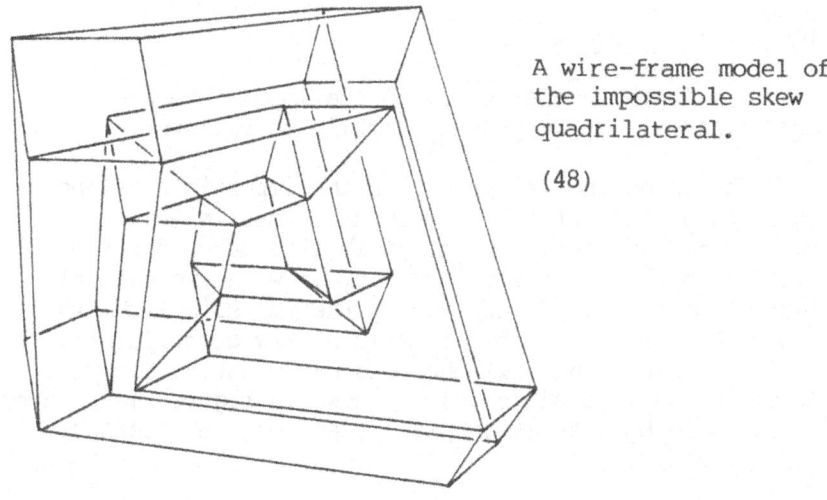

A wire-frame model of
the impossible skew
quadrilateral.

(48)

piece. For the complete model, you will need two pieces of
each handedness. Once assembled, the model is stable in
two orientations: balanced on edge as in figure 52a, or
sitting down, as in figure 52b. The appearance of the
right angle in figure 52b is rather surprising.

The impossible skew quadrilateral works well as a
deceptive 3-d interlocking puzzle. The difficulty is due
to both the unfamiliar angles and the indescribable shape
-- by the time someone is able to reassemble the four
pieces, the form is inevitably different than remembered.
For those not content with the one form, figure 53a shows
an alternate form which can be assembled from the same four
pieces. The "hyper-swastika" has the same symmetry as the
impossible skew quadrilateral. Note the octahedral hole at
the center. Unfortunately, the figure is not stable by
itself since the parts do not interlock. The figure does
not have any 4-d illusionary aspects, either. Figure 53b
shows its counterpart one dimension lower.

We have followed the recipe for the impossible
triangle to the letter, thereby producing its 4-d analog.
It remains to be seen whether it is really impossible. For
all we know, the impossible triangle may work only in 3
dimensions. Or perhaps illusions of this type are so
common in 4-space that it would not be particularly
impressive. Since vision is a perceptual problem, not a
mathematical one, we can't know for sure without asking a
4-d person. Perhaps impossible skew quadrilaterals can be
put out as bait. An intriguing alternative is to program a
simulation of 4-d visual perception.

Impossibility Proof

The depth inconsistency of the impossible skew
quadrilateral can be proved formally as follows.

The impossibility proof for the impossible triangle
involved a cycle of three faces which met in an
inconsistent manner. The impossibility proof for the
impossible skew quad involves a cycle with two elements
instead of three. In explaining the impossibility, the
proper 2-d analog of the impossible skew quadrilateral is
an impossible (non-skew) quadrilateral, not a triangle.
Consider the impossible object shown in figure 54a. Faces
A and B are by assumption planar, and meet at distinct
lines 1 and 2.

By non-degeneracy, faces A and B do not lie in the
same plane. This is immediately impossible, however, since

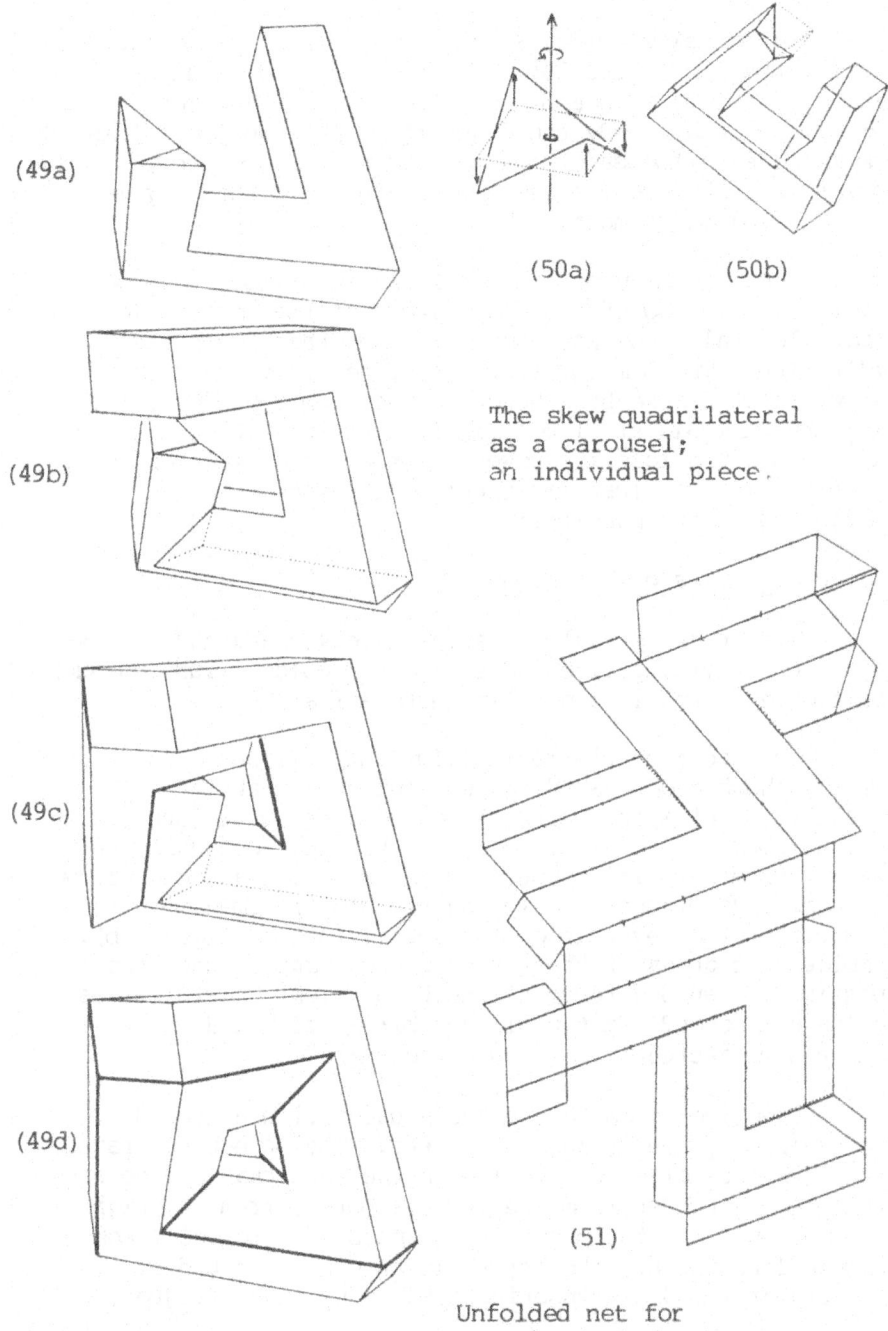

(49a)

(50a)   (50b)

The skew quadrilateral
as a carousel;
an individual piece.

(49b)

(49c)

(49d)

(51)

Piece-by-piece assembly
of the 3-d model.

Unfolded net for
one piece of the
impossible skew quadrilateral.

two distinct planes may meet in one line at most.

For the skew quadrilateral, consider only the two 3-d faces shown in figure 54b. Faces A and B by assumption each lie entirely in a 3-plane and meet at distinct 2-planes 1 and 2. By non-degeneracy, the two faces A and B do not lie in the same 3-plane. This is immediately impossible, however, since two distinct 3-planes may meet in one 2-plane at most.

It may seem surprising that the impossibility proof requires only two of the four pieces of the impossible skew quadrilateral. Why not just omit the other two pieces altogether? The same question could be asked one dimension lower about figure 5b. Here the answer becomes evident: the object would still be technically impossible, but the illusion would lose its power. For an impossible object to be effective, it must first entice the viewer into believing in its possibility.

Different Types of Illusions

What other optical illusions have 4-d analogs? If 4-d eyes work anything like 3-d eyes, then color illusions such as figure 2a should generalize automatically.

What about multi-stable illusions? In our discussion of the rhombic dodecahedron, we mentioned that the tesselation of space using rhombohedra can be interpreted as the image of an array of hypercubes in two equally valid ways. From here it is possible to construct a whole class of ambiguous figures, including the higher-dimensional analogs of the reversible staircase illusion (figure 2b). Ambiguous figures of the figure-ground variety are also possible. Can 3-d tesselations be as richly pictorial as those of Escher? What would 4-d birds, fish and other animals projected into 3-space look like?

Effective impossible objects using figure-ground inconsistency seem harder to extend. The object portrayed in figure 2c relies critically on the fact that a line can represent a border from the left, a border from the right, or a crease. Furthermore, it is impossible to tell from a single line whether the border is angular or rounded. Unfortunately, 4-d boundaries tend to be less ambiguous.

One illusion which does have a straightforward generalization is the endlessly rising staircase (figure 2d; see [Kim] for the 4-d analog). The 4-d endlessly rising staircase is one example of an illusion which

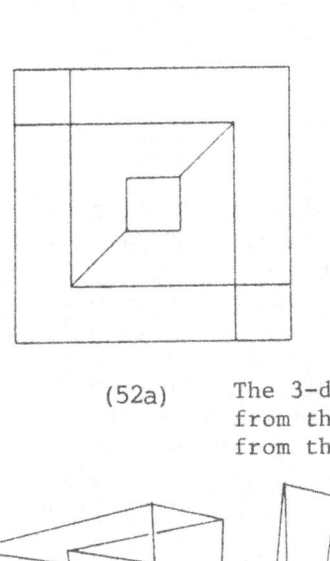

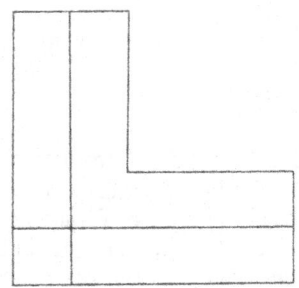

(52a) The 3-d model seen (52b)
from the top and
from the side.

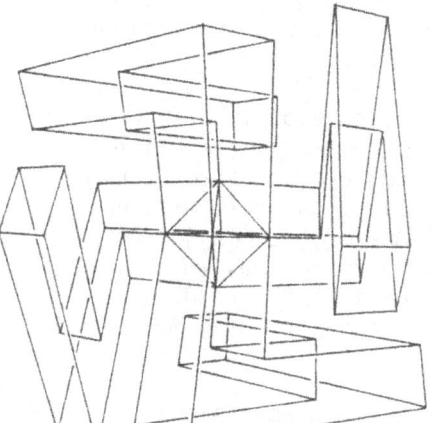

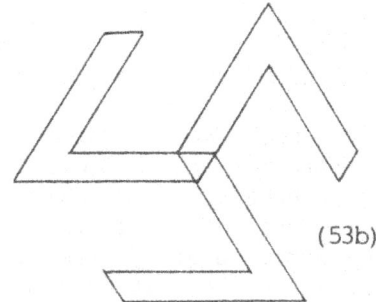

(53b)

The hyperswastika--an
alternate shape from
the same pieces; the
2-d analog.

(53a)

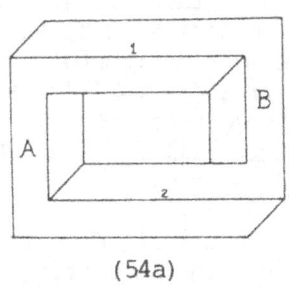

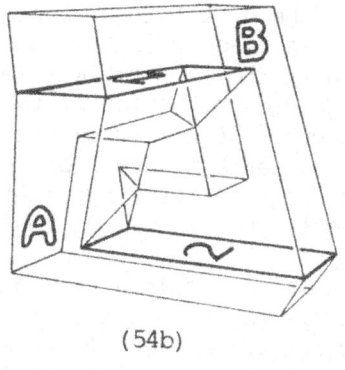

(54a)

(54b)

An impossible (non-skew)
quadrilateral.

The impossible skew
quadrilateral labeled for
an impossibility proof.

actually works better in 4 (or more) dimensions than it does in 3.

Conclusions

Summary Remarks

In this article, we have constructed a 4-d optical illusion. Mathematically the construction is simple. But optical illusions are figments of perception, not of mathematics. Designing an illusion requires knowledge of vision (the psychology of expanding 2 dimensions into 3) as well as of drawing (the mathematics of projecting 3 dimensions into 2). Since we have chosen to construct optical illusions in a domain where our standard intuitions fail (4-space), we have had to clarify the assumptions underlying vision and dimension. Studying the impossible skew quadrilateral brings 3 dimensional space and ordinary optical illusions into sharper focus.

The most important step in construction is understanding the rhombic dodecahedron and its relation to the hypercube. The impossible triangle appears to be made of cubes joined face to face. Similarly, the impossible skew quadrilateral appears to be made of 3-cubes joined 2-face to 2-face. It is also necessary to know which lines are hidden when two cubes are joined in order to draw aggregates of cubes. Hidden line removal becomes most complicated at corners.

When I originally pondered the impossible skew quadrilateral, most of the work went into drawing the structure of a corner. At that time I was still convinced that the global form should be a tetrahedron. Everything fell in place when Philip Wadler suggested that the global form of the impossible triangle could be derived from the isometric projection of the cube. The most delightful aspect of generalizing the impossible triangle to 4-space is the certain intuition that such an object can be logically derived, but that its structure will nonetheless be quite non-intuitive to 3-d eyes.

The impossible skew quadrilateral as presented in this article can be modified in several ways. The isometric projection of the cube was used only as a simplifying assumption; impossible objects may equally well be based on non-isometric projections as long as they are not degenerate. We could also have made the beams in the impossible skew quadrilateral narrower, creating a larger central hole. In the present model the beams turn so

quickly they are not seen distinctly as hexagonal prisms.
Only a unit distance separates the corner rhombic
dodecahedra. One unit less and the hole would have
collapsed to zero.

Open Problems

The impossible skew quadrilateral is the 4-d analog of
a 3-d illusion: the impossible triangle. Are there
optical illusions in four or more dimensions which do not
have lower-dimensional analogs? Such an illusion would
probably exploit techniques specific to 4-space. There may
also be illusions which are distinct in higher dimensions,
but map down onto the same 3-d equivalent. For instance,
we have described the Star of David as the 2-d analog of
both the stella octangula and the stellated rhombic
dodecahedron.

One of the original hopes in constructing the
impossible skew quadrilateral was that a 3-d person would
somehow be able to directly perceive the 4-d illusion,
perhaps without realizing precisely how. We would then
have a direct bridge to the visualization of 4-space.
People with unusually refined 4-d intuitions are described
in [Coxeter], summarized in [Muses]. Perhaps such a person
would be able to appreciate the impossible skew
quadrilateral. But if people cannot perceive the 4-d
illusion directly, then perhaps computer programs can.

It is interesting to speculate on the requirements for
a 4-d visualizer. First we might want to assume, as we
have been doing all along, that vision in other dimensions
is very similar to our own 3-d vision. As already
mentioned, vision cannot be separated from a knowledge of
which shapes are likely to occur. Even limiting ourselves
to straight-line drawings as we have been doing, facts
about the world, such as the presence of gravity, are
unavoidable. Vision would be quite different if there were
no gravity, there were eyes at the ends of our fingers, or
the speed of light were much slower.

Throughout this article we have compared the jump from
3 to 4 dimensions with the jump from 2 to 3 dimensions.
Would such parallelism work equally well if we were going
from 4 to 5 dimensions? For the construction of the
impossible triangle and its analogs, the answer is yes —
the same idea works in all higher dimensions. Once the
concepts are put in the proper form, one need only
substitute the correct dimension to produce the equivalent
object. Cubes, beams, corners and loops all generalize

directly.

This suggests that it might be possible to parametrize visual perception with respect to dimension. Think of a general visualizer into which you plug the number of dimensions. One of the most intriguing properties which should carry over into 4-space is our apparent inability to visualize higher-dimensional space. A 4-d creature would probably be just as uncomfortable with the idea of 5-space as we are with 4.

However, many ideas simply don't carry over into other dimensions. After all, the impossible triangle does not work in 2 dimensions. One can add a proviso covering the special case, but there are many other dimension-specific properties to account for. For instance, knot-tying is possible only in 3 or more dimensions, while the length of the spatial diagonal of a unit n-cube is integral only in square dimensions. Enumerating the quirks of all dimensions can be compared with listing the properties of all the positive integers. Do new ideas emerge each time you go up a dimension? If so, is it possible to predict what they will be?

Escher

In 1962, M. C. Escher completed a lithograph called "Waterfall" (figure 55a, [Ernst]). "Waterfall" appears to be a 3-d structure composed of rectangular viaducts. But as with all of Escher's works, appearances can be deceiving — no consistent 3-d model of "Waterfall" can be built. It exists only on paper. Escher worked through several variations of the illusion before settling on the idea of using flowing water to emphasize the absurdity of the false connection [Ernst]. "Waterfall" was directly based on the work of the mathematician R. Penrose, who developed the optical illusion shown in figure 1a ("the impossible triangle").

The underlying structure of "Waterfall" is a stack of three impossible triangles (figure 55b), extracted from the impossible triangular grid (figure 55c). Each triangle individually is impossible. If we leave out the vertical pillars, keeping only the diagonal aqueducts, the resulting structure appears to flow slightly downhill off into the horizon in a perfectly reasonable manner. If we continue the water flow downhill off the lip of the most distant aqueduct, it should fall into a subterranean pool. In this picture, however, it connects to the original reservoir, completing the circuit. Escher takes advantage of the

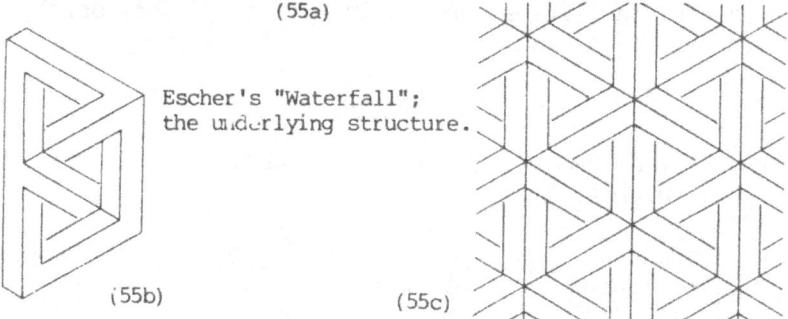

(55a)

Escher's "Waterfall";
the underlying structure.

(55b)   (55c)

Lithograph by M. C. Escher, "Waterfall," reproduced by
permission of the Escher Foundation, Haags Gemeentemuseum,
The Hague.

ambiguities of representing 3 dimensions on 2-d paper to create a false connection.

But there's more to the presentation of "Waterfall" than just a triad of impossible triangles. "Waterfall" does not use any perspective -- beams do not shrink as they recede. Instead, 3-dimensionality is emphasized by building up richly 3-d textures, such as the polyhedra atop the towers, the terraces in the background, and the detailed brickwork. Escher further emphasizes the fantastic nature of his waterfall by surrounding it with decidedly nonparadoxical spectators. One is tempted to consider the 4-d natures of polyhedra, contour plowing, bricklaying, waterspheres, clothesplanes, and hyperfungus. What is the 4-d analog of "Waterfall"?

Acknowledgements

I would like to acknowledge the help of Thomas Cover at Stanford University, who prompted me to study 4-d optical illusions in the first place; Philip Wadler who helped break some crucial mental blocks; Doris Schattschnieder, who encouraged publication and connected me with Roger Penrose; Roger Penrose for sharing the same interest in the mathematically absurd; Bruce Baumgart, author of the polyhedron-displaying program "Geomed", which was invaluable in preparing the illustrations; the Stanford Artificial Intelligence Laboratory and the Institute for Mathematical Studies in the Social Sciences at Stanford for use of their computer text-editing facilities; Douglas Hofstadter and Roger Shepard for their valuable comments and suggestions; and for the present revision: Logicon (San Pedro, California) for the use of their computer text-editing facilities, and David Brisson for editorial assistance.

Bibliography

Attneave, Fred. "Multistability in Perception." Scientific American, December 1971.

Cowan, Thaddeus M. The Theory of Braids and the Analysis of Impossible Figures. Journal of Mathematical Psychology [11], 190-212 (1974).

Coxeter, H. S. M. Regular Polytopes. Dover Publications, 1973.

Critchlow, Keith. Order in Space. The Viking Press, Inc., 1970.

Ernst, Bruno. The Magic Mirror of M. C. Escher. Random House, N.Y., 1976.

Gregory, R. L. The Intelligent Eye. McGraw-Hill Book Company, 1970.

Hilbert, D., and Cohn Vossen, S. Geometry and the Imagination. Chelsea Publishing Company, 1952.

Holden, Alan. Shapes, space, and symmetry.

Huffman, D. A. "Impossible objects as nonsense sentences." Pages 295-323 in
B. Meltzer and D. Michie (eds.). Machine Intelligence 6. Edinburgh University Press, 1971.

Kim, Scott. "The endlessly rising staircase in 4 dimensions". To be published.

Muses, Charles, and Young, Arthur, eds. Consciousness and Reality. Outerbridge & Lazard, Inc., distributed by E. P. Hutton & Co., 1972.

Penrose, L. S., and Penrose, R. "Impossible objects: A special type of illusion." British Journal of Psychology [49], 31 (1958).

Penrose, Roger. Personal correspondence. October, 1976.

Winston, Patrick Henry, ed. The Psychology of Computer Vision. McGraw-Hill Book Company, 1975.